ASTRONOMY

A Wm. C. Brown Publishers'
Reader in Astronomy

ASTRONOMY

A Wm. C. Brown Publishers'
Reader in Astronomy

Kalmbach Publishing Company

David Dathe
Alverno College

Wm. C. Brown Publishers
Dubuque, Iowa • Melbourne, Australia • Oxford, England

Book Team

Editor *Jane Ducham*
Developmental Editor *Thomas W. Riley*
Production Editor *Audrey Reiter*
Designer *Jeff Storm*
Photo Editor *Carrie Burger*

 Wm. C. Brown Publishers
A Division of Wm. C. Brown Communications,

Vice President and General Manager *Beverly Kolz*
Vice President, Publisher *Earl McPeek*
Executive Editor *Jeffrey L. Hahn*
Vice President, Director of Sales and Marketing *Virginia S. Moffat*
National Sales Manager *Douglas J. DiNardo*
Marketing Manager *Amy Halloran*
Advertising Manager *Janelle Keeffer*
Director of Production *Colleen A. Yonda*
Publishing Services Manager *Karen J. Slaght*
Permissions/Records Manager *Connie Allendorf*

 Wm. C. Brown Communications, Inc.

President and Chief Executive Officer *G. Franklin Lewis*
Corporate Senior Vice President, President of WCB Manufacturing *Roger Meyer*
Corporate Senior Vice President and Chief Financial Officer *Robert Chesterman*

Cover photo: whirlpool galaxy, M151; courtesy of the National Optical
Astronomy Observatories/N. A. Sharp

A Times Mirror Company

Library of Congress Catalog Card Number: 93–75017

ISBN 0–697–24299–4

Printed in the United States of America by Wm. C. Brown Communications, Inc.,
2460 Kerper Boulevard, Dubuque, IA 52001

10 9 8 7 6 5 4 3 2 1

Contents

PART SEVEN: COSMOLOGY

FROM THE PUBLISHER

Wm. C. Brown Publishers is pleased to offer our customers this exceptional astronomy reader. *ASTRONOMY: A Wm. C. Brown Publishers' Reader in Astronomy* is yet another successful joint venture with Kalmbach Publishing Company, publisher of ASTRONOMY, a monthly magazine.

ASTRONOMY is one of the leading astronomy magazines available to professional and amateur astronomers today. The excellent visuals, wide array of topics and superb contributors made this magazine a logical resource when planning an effective astronomy reader. Wm. C. Brown Publishers is proud to offer you the opportunity to use and enjoy the best of ASTRONOMY magazine. We have carefully selected twenty of the most effective and interesting articles published. We encourage professors to order this reader for your students and to coordinate it with your introductory astronomy course. To assist professors in doing this, we also provide an Instructor's Manual, prepared by David Dathe of Alverno College. This resource provides assistance in incorporating the reader into the course with suggestions for use, quizzes and summaries.

Wm. C. Brown Publishers would like to thank Robert A. Maas, Rhoda Sherwood, Tracy Staedter, Julie Sherwin, Robert Burnham and the members of the Production Department at ASTRONOMY magazine for their professional assistance and the ease with which this project came together.

We trust you will enjoy reading and using this exciting and timely reader. We would most certainly appreciate hearing from you about what you think of *ASTRONOMY: A Wm. C. Brown Publishers' Reader in Astronomy.*

Jane T. Ducham
Astronomy Editor

From the Editors
of ASTRONOMY
Magazine

How to Buy
Your First
Telescope

Buying Your First

Your Window on the Universe

It is quiet. The Sun has set, and with it go the worries of the day. The western horizon is tinged with the deep orange of twilight. Above the band of sunset color the sky fades into a deep, clear blue, punctuated by a few bright planets. Soon they will be joined by hundreds of stars. Though people describe this time as a curtain of night falling, you see it differently — as a curtain rising. You know the coming darkness will actually reveal a wide new stage on which the drama of the universe plays every clear night.

You have waited for this night. With the anticipation of an explorer about to set foot on a new continent or a new world, you set up your telescope. Under the twilit sky you check its fittings, orient its mount, insert a favorite eyepiece, and wait for the night to bring forth your list of chosen targets. Tonight it's a journey out to clusters of newborn stars along the Milky Way, then to a glowing cloud of interstellar debris cast off by an ancient supernova explosion. The journey will end with a glimpse of spiraling whirlpools of stars, galaxies so distant that their light — the light you'll capture tonight — has been traveling since the time of the dinosaurs. All this is within reach of your telescope.

When you own a telescope, each evening like this is a welcome time to escape beyond the confines of the world. It is an opportunity to extend your vision into the depths of space, to explore cosmic vistas few others have the chance to see. But the privilege is yours every clear night. Your telescope is your personal window on the universe.

Telescope

A Telescope Buyer's Top 30 Questions

Buying your first telescope can seem like a complicated affair. There are many models to choose from and many technical terms to contend with. To help you make the right choice, here are answers to 30 of the most-asked questions we get from prospective telescope buyers. We trust you'll find answers to your questions among them. If not, please call us at ASTRONOMY magazine [(414) 796-8776, Monday to Friday, 8:30 a.m. to 5:00 p.m. Central Time] or see your local telescope dealer.

TRIO OF TELESCOPES: REFRACTORS use lenses to collect and focus light. They combine ruggedness with crisp images.

A NEWTONIAN REFLECTOR uses a concave mirror at the bottom of the tube, which reflects light to a small, flat secondary mirror which then deflects the light to the eyepiece at the side of the tube.

In a SCHMIDT-CASSEGRAIN, a concave mirror reflects light to a convex secondary mirror which in turn sends the light back down to a focus at the rear of the telescope. A thin corrector plate at the front of the tube compensates for an optical distortion called spherical aberration.

1 How much does the telescope magnify?

Beware of any telescope advertised as "500x" or "high-power." Some manufacturers make it sound as if the more magnification a telescope offers, the better it is. This is not true. Contrary to the claims of department-store catalogs, *magnification is not the most important quality.* Any telescope can be made to magnify any amount. However, the highest power that will still give you a clear view is about 50x per inch of aperture, making the upper limit for a 3-inch telescope 150x and for a 4-inch telescope 200x. Beyond this limit, the image will be faint, fuzzy, and disappointing.

2 How, then, do I select the best telescope?

The key characteristic of a telescope is its aperture — the diameter of the main lens or mirror. The larger the aperture, the more light the telescope gathers and focuses into the image. This in turn makes for a brighter, and usually clearer, image. Brighter images make it easier to see faint objects like nebulae and galaxies.

3 Which are better — refractors or reflectors?

A refractor uses a lens mounted at the front of the telescope to gather and focus light. A reflector uses a concave mirror mounted at the back of the telescope. Both work well; each has its advantages. Reflectors generally offer more aperture for the money. (A 4-inch reflector costs $400 to $500; 4-inch refractors start at $1,000 or more.) However, refractors usually

REFRACTOR

Eyepiece Holder

Lens

NEWTONIAN REFLECTOR

Eyepiece Holder

Primary Mirror

Secondary Mirror

SCHMIDT-CASSEGRAIN

Corrector Plate

Eyepiece Holder

Secondary Mirror

Primary Mirror

What Can I See?

What you can see through the eyepiece of a good telescope is enough to keep you exploring the sky for many years. Here's a bit of advice: when you unpack your first telescope and set it in the backyard on its premiere night under the stars, the first thing you should look at is the Moon. You won't need high power — 50x will be just fine. Even at this low power, you'll be amazed at the view — we guarantee it.

After observing the Moon, the first planet you should look at is either Jupiter or Saturn. (This winter or spring try for Jupiter; in summer or autumn try Saturn — the monthly "Sky Almanac" section of ASTRONOMY will tell you where to find them.)

When you aim your scope at Saturn, be prepared for a remarkable sight. Most people utter an exclamation of "Wow!" when they first see Saturn and its picture-perfect rings.

Even a small 60mm-aperture telescope will reveal the cloud belts on Jupiter. Over the course of several nights, you'll see Jupiter's four large moons as bright dots shuttling back and forth from one side of Jupiter to the other.

A telescope also allows you to follow the changing phases of Mercury and Venus, and to watch Mars grow and shrink in size as we approach and then recede from the Red Planet every two years. And, yes, Mars really does look orange-red. When it is closest to Earth, a magnification of 100x to 200x will show its reddish disk, dark surface markings, and famous polar caps. You can continue to tour the solar system by tracking down Uranus and Neptune, although even in large telescopes they appear only as tiny blue-green dots. Starlike Pluto is within reach of backyard scopes but requires at least an 8-inch-aperture instrument and dark skies.

Beyond the solar system there are hundreds of star clusters, nebulae, and galaxies within reach of 60mm- to 80mm-aperture scopes. Larger scopes reveal these deep-sky objects in even more detail and bring thousands more deep-sky wonders within reach. But don't expect to see the colors you see in photographs — the colors of nebulae and galaxies are so faint they show up only in long-exposure photographs. Most deep-sky objects appear as misty, gray patches of light. However, some stars show colors your unaided eye cannot see, and many stars that appear as single to the naked eye appear split into two or more stars through a telescope — they are systems of alien suns orbiting each other.

If you want an idea of how far you can see with a telescope, you can actually explore the most distant reaches of the cosmos by hunting down elusive galaxies. Because of their immense distances, these islands of stars are typically very faint. Nevertheless, from a dark rural site a 3- or 4-inch telescope will show many of these ghostly spots of light, enabling you to probe tens of millions of light-years into space. All with a telescope costing no more than a round-trip air fare to Europe. It's a bargain price for a ticket to the stars.

2-inch telescope (50x)

4-inch telescope (125x)

12-inch telescope (275x)

**Whirlpool Galaxy
(8-inch telescope)**

provide slightly sharper images than reflectors of similar aperture. Amateur astronomers who like to view fine details on planets often prefer refractors; those who like to look at faint deep-sky objects use reflectors. For most first-time buyers on a budget, either an 80mm refractor or a 4.5-inch reflector is a good choice. Both cost $400 to $600 and have comparable performance.

4 What are Schmidt-Cassegrain telescopes?

A third type of telescope system, called a catadioptric, uses a combination of mirrors and a refractive corrector lens at the front. The most popular of these hybrid models is the 8-inch Schmidt-Cassegrain (prices start at $1,200). It folds a long focal length into a compact tube, making this type of telescope very portable and convenient to use for its aperture. It is also a good general purpose telescope suitable for observing all classes of celestial targets.

The two main manufacturers of this type of telescope are Celestron International and Meade Instruments Corporation. These two companies are competitive and offer a similar range of Schmidt-Cassegrains, from basic no-frills units to feature-laden computer-controlled models. Which company makes better telescopes? Over the years we have found that in optical and mechanical quality, both are nearly equal.

5 What are apochromatic refractors?

One of the principal problems with conventional refractors over 80mm aperture is spurious color around bright objects caused by the inability of the lens to bring all colors to the same point of focus. To greatly reduce this "chromatic aberration," manufacturers have introduced refractors that use 3-element lens systems or special fluorite or "ED" lenses. Called apochromatics, these high-end refractors are among the finest optical systems you can buy and have become popular with telescope aficionados. Models are available from Astro-Physics, Celestron, Takahashi, Tele Vue, Vernonscope/aus Jena, and Vixen. However, a 4-inch "apo" refractor can cost $2,500 to $5,000, more than most people wish to spend on a first telescope.

6 How much more will I see with a bigger telescope?

Bigger telescopes can show fainter objects and resolve finer details in bright objects. For example, a 2.4-inch (60mm) refractor will easily show the cloud belts of Jupiter, a 4-inch will show structure within the cloud belts, and an 8-inch will resolve even smaller details. A 4-inch will show globular star clusters as fuzzy-edged spheres of light, a 6-inch will resolve many globulars into myriad faint stars, and a 12-inch will provide views of these clusters that surpass any photograph. While a 4-inch will reveal a spiral galaxy as a round glow, an 8-inch will begin to reveal the galaxy's spiral arms.

A QUARTET OF MOUNTS:
ALT-AZIMUTH mounts swing left-to-right and up-and-down. They are inexpensive but cannot automatically track the stars.

The DOBSONIAN mount is an alt-azimuth design often used on large reflectors.

The GERMAN EQUATORIAL mount can track the stars with a single motion. It is used with long-tube telescopes such as refractors. Motor drives are often options.

The EQUATORIAL FORK mount can also track the stars and is used on short-tube telescopes like Schmidt-Cassegrains. Motor drives are built into the base.

ALT-AZIMUTH MOUNT

GERMAN EQUATORIAL MOUNT

DOBSONIAN MOUNT

FORK MOUNT

REFRACTORS ARE POPULAR beginners' telescopes. The better grade of small 60mm models such as the Meade #290 (bottom) can get you started. But we recommend 80mm refractors, such as the alt-azimuth-mounted Celestron FirstScope 80 (below) or the similar Meade #312, as better choices.

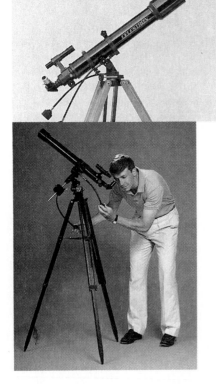

HIGH-END REFRACTORS such as the Astro-Physics 4-inch apochromat (below) offer top-notch optics but at premium prices. Other competitive models are available from Celestron, Takahashi, and Tele Vue (their popular "Genesis").

7 Will I be happy with a smaller scope?

The fact that bigger telescopes usually show more details and fainter objects leads many people to believe small scopes aren't worth buying. But even an 80mm refractor can show you enough of the universe to keep you entertained for years. For many people it's all the telescope they ever need.

We warn people against buying a telescope that is too large — yes, there is such a thing. A big telescope, though exciting at first, can quickly become a burden to carry out to the yard or car and to set up. The best telescope is not the biggest, or even the one with the best optics, but *the one that you will use most often*. Portability and convenience are factors to consider when selecting a telescope that you'll have fun using.

8 I live in the city (country) where the skies are terrible (great). What type of telescope is best for me?

A large-aperture telescope can be useful at any site, but faint deep-sky objects (the kind big scopes are well-suited for) won't show up well under urban skies, no matter what size the scope. City observers often spend more time looking at the Moon and planets, for which a 3- to 8-inch telescope is sufficient. Telescopes in that size range are also very portable, an important consideration for city observers who need to transport their scopes to better skies. For most buyers, we feel that a 5-inch refractor, a 6-inch equatorial reflector, an 8-inch Schmidt-Cassegrain, or a 10-inch Dobsonian reflector (see question 16) are the largest telescopes of their types that are conveniently portable. Only if you live under dark skies, or really don't mind lugging a big, heavy scope around, should you consider anything larger for a first telescope.

9 What does focal length mean? And f/ratio?

The focal length of a telescope is the length of the light path from the main lens or mirror to the eyepiece. In most refractors or reflectors, the focal length is roughly the length of the tube. In telescopes such as Schmidt-Cassegrains in which the light path is bounced back and forth inside the tube several times, the length of the tube is much shorter than the focal length. Focal lengths of telescopes, like camera lenses, are usually measured in millimeters.

The f/ratio of a telescope is the focal length divided by the aperture. For example, a 100mm-aperture telescope with a 900mm focal length is an f/9 telescope. A 200mm telescope (an 8-inch) with a focal length of 1,800mm is also an f/9.

10 What focal length is best?

The focal length of the telescope is not a critical specification. Telescopes with shorter focal lengths (400 to 700mm) will give lower powers and wider fields of view with a given eyepiece than telescopes with moderate (800 to 1,200mm) or long focal lengths (1,300 to 3,000mm). For this reason, short focal lengths are often preferred for wide-angle viewing of deep-sky targets and Milky Way starfields. On the other hand, a long-focal-length scope will give a higher power with any given eyepiece. Since the planets require higher powers (100x to 200x), planetary fans often prefer long-focal-length scopes. But with the use of the appropriate eyepieces, most telescopes can be used at both low and high powers.

11 How can a telescope be "fast" or "slow"?

Sometimes manufacturers give the impression that a "faster" telescope (one with an f/ratio of f/4 to f/6) is better than a "slow" telescope (f/7 to f/16). After all, in many situations faster *is* better. But in this case it isn't. The term "faster" comes from photography, where an f/4 lens will record an image with a faster exposure time than an f/16 lens will. And for those intending to take long-exposure photos through a telescope, faster scopes can be better. But when you look through a telescope, a faster telescope does not give brighter images than a slower scope. For example, as long as both are operating at the same power, the image in an 8-inch f/6 telescope will appear as bright as the image in an 8-inch f/10 scope. The difference is that with the same eyepiece the f/6 telescope will give a lower power and a wider field of view than the f/10 will, making faster scopes preferred for deep-sky observing where wide fields are desirable.

12 What is an eyepiece for?

Eyepieces allow you to change magnification of the image formed by the lens or mirror. To determine the magnification an eyepiece gives, divide the focal length of the telescope by the focal length of the eyepiece. For example, a 25mm focal-length eyepiece used on a 2,000mm focal-length scope (such as an 8-inch f/10 scope) will give 2000/25 = 80x. The same eyepiece used on a 1,600mm scope (such as an 8-inch f/8) will give 64x.

13 What is the little scope for?

The little scope you see on many telescopes is a finderscope. It is an essential accessory. It provides a low power (5x to 8x) and a wide field (3° to 5°) and allows you to aim the telescope easily and center it on bright planets and stars. Without a finderscope, locating even the Moon can be difficult.

14 Why are images in telescopes upside-down?

All astronomical telescopes present images that are either upside-down or flipped left-to-right as in a mirror. To flip the image right-side up would require extra lenses in the light path that would dim the view of already faint astronomical objects or add imperfections like flares and ghost images.

15 Which is better — an alt-azimuth or equatorial mount?

Alt-azimuth mounts use simple up-down (altitude) and side-to-side (azimuth) motions to aim the telescope. The best of these mounts are equipped with slow-motion controls that allow you to make fine adjustments to the position of the scope. However, alt-az mounts cannot automatically follow the stars as they arc across the sky from east to west.

An equatorial mount is more complex. It *can* follow the stars across the sky with a single motion around one axis. If the telescope is equipped with a motor, the telescope will automatically track the stars. This is a nice feature because at magnifications of 100x or more, the apparent motion of the sky will cause objects to drift out of the field of view in less than a

REFLECTORS OFFER THE MOST aperture for the money. The Celestron C4.5 (above) is one of the best of many 4.5-inch reflectors on the market. A 6-inch reflector, such as the Parks 6-inch f/8 Precision (below) costs more but makes a superb all-purpose telescope.

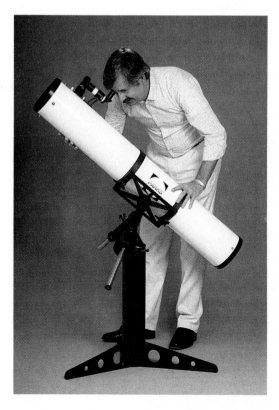

minute. Having to recenter the image constantly can be distracting and inconvenient, and can introduce vibration that shakes the image.

16 What about Dobsonian telescopes?

A Dobsonian is a Newtonian reflector. Its unique feature is a simple wooden alt-azimuth mount that rides on Teflon pads. The philosophy of the popularizer of this type of telescope, John Dobson, was to keep the scope easy-to-build and low-cost. The design also lends itself to relatively large apertures. Dobsonians cannot track the stars automatically, but their motions are very smooth — it's easy to nudge the scope every so often to recenter the object.

As of 1991, only one manufacturer was selling low-cost Dobsonian telescopes — Coulter Optical Company. Their models offer big aperture for very little money (for example, a 10-inch reflector for under $400). People often wonder if there's a catch. There are a few: Coulter scopes are in such demand that you may have to wait several months to a year for delivery. You must order directly from the factory and pay shipping costs from their plant in California (which may add up to $100 to the cost). You will also need to add a finderscope. The fit and finish of the scopes is nothing fancy — the mounts are painted chipboard, the tubes cardboard. Our opinion? *For the money,* the quality of construction and optics can't be beat. The 8- and 10-inch Coulter models make good starter scopes. The 13- and 17-inch models are best for die-hard deep-sky observers.

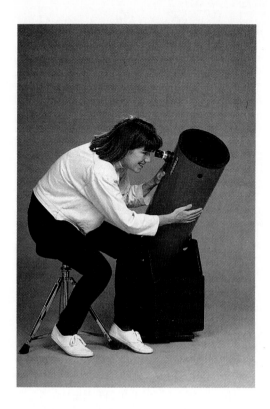

OFTEN CALLED "LIGHT BUCKETS," Dobsonian-mounted reflectors, such as this Coulter Odyssey 8, offer lots of aperture on an ultra-simple but steady mount. Deep-sky explorers willing to sacrifice features and finish for aperture and low price love Dobsonians.

17 I've heard you can make your own telescope.

The Dobsonian design lends itself to do-it-yourselfers. Plywood for the mount and a cardboard tube like those used for concrete forms are the main ingredients. Few people make their own mirrors these days. It can be done, but ready-made mirrors from suppliers like Coulter, Meade, and Parks don't cost much more than mirror-making kits. You'll also need a focuser and cells to hold the main mirror and the small secondary mirror. For more information about telescope making, see the book *Build Your Own Telescope* by Richard Berry, available from Kalmbach Publishing.

18 What accessories do I need?

Some telescopes come with only one eyepiece. Additional eyepieces for higher and lower powers are the first accessories most first-telescope owners need to buy. An accessory called a Barlow lens can double or triple the power of each eyepiece, but the best Barlows (the only ones worth buying, trust us!) cost $80 to $100.

Colored filters can enhance views of the planets slightly, but the difference is subtle. They are not essential. Nebula or light-pollution filters can improve views of some deep-sky objects like emission and planetary nebulae, but they do little to improve star clusters and galaxies. Contrary to what many beginning backyard astronomers believe, these filters are not a cure-all for light-polluted skies.

Computerized digital readouts to aid in finding objects have become popular telescope accessories in recent years. They work well but are luxury options for those that can afford their $500 to $1,000 price tags.

19 Are enhanced coatings worth the extra cost?

Some telescopes are offered with special lens or mirror coatings as optional extras (Celestron's Starbright™ and Meade's MCOG, for example). These increase the light transmission, yielding images up to 15 percent brighter. They are definitely worth the extra expense.

20 Can I use setting circles to find things?

Many equatorial mounts are equipped with graduated dials called setting circles. Theoretically, these allow you to find objects by moving the telescope so that the circles' readings match the celestial coordinates (called right ascension and declination) of the object you're looking for. However, in our experience we have rarely seen a novice amateur astronomer (nor many experienced ones!) who have been able to make effective use of setting circles. Poor alignment of the telescope mount, improperly calibrated circles, and imprecise circle scales usually combine to make circle readings inaccurate. The best method to find celestial targets is to hop from star to star using a good star chart as your guide. Plan on buying such a star chart as an essential accessory.

21 Can I take pictures with this scope?

Anything you see through a telescope can be photographed, but most objects require exposures of several seconds to an hour or more. Keeping the object perfectly positioned on the film during that time requires a solid equatorial mount and a motor drive. These are essential features if you intend to do astrophotography.

22 Can I use a spotting scope for astronomy?

Some spotting scopes (such as those sold for birding) have only a fixed-power eyepiece or a variable zoom eyepiece. These models are not well suited for astronomy. Other models use interchangeable eyepieces but must be placed on a solid camera tripod. Because they lack fine slow-motion controls, camera tripods are difficult to aim precisely, a problem at high power.

23 But I also want to use my telescope for nature observing.

If your interests mix astronomical and terrestrial viewing, we suggest an 80mm or 4-inch refractor or a small 4-inch Schmidt-Cassegrain. Don't buy a Newtonian reflector — the position of the eyepiece makes a Newtonian awkward for use as a spotting scope.

Performance Specs

Aperture (inches)	(mm)	Faintest Star[1]	Resolution[2] (arcseconds)	Highest Usable Power[2]
2.4	60	11.6	2.0	120x
3.1	80	12.2	1.5	160x
4	100	12.7	1.2	200x
5	125	13.2	1.0	250x
6	150	13.6	0.8	300x
8	200	14.2	0.6	400x
10	250	14.7	0.5	500x
12.5	320	15.2	0.4	600x
14	355	15.4	(0.33)	(600x)
16	400	15.7	(0.3)	(600x)
17.5	445	15.9	(0.27)	(600x)

[1]The faintest star visible for that size telescope using the magnitude scale, where a mag. 6 star is the faintest the unaided eye can see, mag. 11 is 100 times fainter, and mag. 16 is 100 times fainter again. Using superb optics under ideal conditions, experienced observers can often exceed these values by a magnitude or more.

[2]Under most atmospheric conditions, even in large telescopes, magnifications over 500x to 600x and resolutions better than 0.4 to 0.5 seconds of arc are rarely possible.

PORTABILITY IS A CRITICAL FACTOR for many people. Compact telescopes such as the Meade 2045 4-inch Schmidt-Cassegrain (above) and the Edmund Astroscan (below) are easy to carry and set up anywhere with little fuss. The 4.2-inch Astroscan reflector, a fine starter scope, is shown sitting on a homemade plastic-pipe tripod.

Telescope Pros and Cons

	The Pros	The Cons
Achromatic Refractors (60mm to 5-inch)	Economical in smaller sizes; rugged; portable; easy to aim; usually provide sharp images.	Small apertures have limited light-gathering power; larger apertures exhibit chromatic aberration.
Apochromatic Refractors (3- to 7-inch)	Provide high-quality images near perfection; excellent for lunar and planetary viewing; fast models good for wide-field, deep-sky viewing and photography.	Relatively expensive for the aperture; light-gathering power cannot compete with that of larger reflectors.
Equatorial Newtonian (4- to 18-inch)	Large aperture for the money; small sizes are excellent scopes for serious beginner. In f/6 to f/8 designs, they are good all-purpose scopes.	Can be very bulky and heavy in sizes over 8 inches; mirrors require adjustment; mirror surfaces are exposed and can get dirty.
Dobsonian (8- to 20-inch)	Biggest aperture for least money; portable for the aperture; superb for deep-sky observing; easy to set up (no polar alignment); great for dark-sky sites.	Optical quality in low-cost models is a compromise; mount does not track the stars; mirror collimation critical in fast f/ratio models.
Schmidt-Cassegrain (4- to 11-inch)	Very portable for an equatorially mounted scope; easy to set up and aim; adaptable for astro-photography; expandable systems with many accessories; good general-purpose telescopes.	Outperformed by specialized telescopes for planetary (refractors and long-focus Newtonians) and deep-sky viewing (large Dobsonians); corrector plates attract dew.

24 How much should I have to spend?

We feel that $400 to $500 is the minimum for a quality starter scope such as an 80mm refractor or 4.5-inch reflector. The next step up is to a 6-inch equatorially mounted reflector (such as the models from Celestron, Meade, or Parks Optical). These sell for $600 to $900. The next level up that many first-time buyers consider is an 8-inch Schmidt-Cassegrain ($1,200 to $2,500). No-frills Dobsonian reflectors defy these price/aperture categories by offering much more aperture for the money (see question 16).

25 What does "$1/20$th-wave optics" mean?

The deviation of an optical surface (lens or mirror) from the ideal shape is often stated as a fraction of a wavelength of light. The smaller the fraction, the better the optics and the sharper the image. However, to be meaningful for a complete telescope, this deviation figure should be provided for the final wavefront reaching the eye, not just for individual lenses or mirrors. When measured in this manner, a telescope with a total error on the final wavefront of $1/4$ wave is good, $1/8$ or $1/10$ wave is excellent, and $1/20$ wave is outstanding though seldom achieved. Manufacturers have no agreed-upon standard for measuring these values — one company's $1/20$ wave may be equivalent to another company's $1/10$ wave.

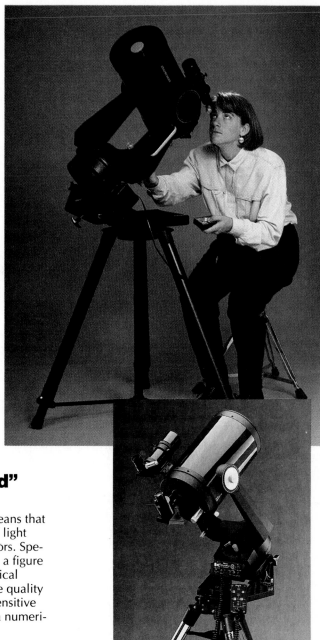

26 What does "diffraction limited" mean?

This is another freely used term in telescope advertising. It means that the optics are so good they are limited only by the wave nature of light and not by any flaws in the surface accuracy of the lenses or mirrors. Specifically, it means the final wavefront error is better than $1/4$ wave, a figure known as Rayleigh's Criterion. Few manufacturers have the technical equipment to quantitatively support this claim. Most test telescope quality by ensuring units form good star images. Although this *is* a very sensitive test that will detect small flaws in the optics, it cannot guarantee a numerical specification like $1/4$ wave.

27 Where can I buy a good telescope?

We suggest shopping at a local telescope dealer if there is one near you (check the Yellow Pages under "Telescopes"). Good dealers will check each scope they sell, provide good service, answer your technical questions, and perhaps allow you to take home a scope on a trial basis. You can at least see what you're getting before you buy it. This peace of mind is worth any extra cost involved.

Mail-order companies that specialize in astronomy products can also offer personal service (over the phone) and money-back guarantees of satisfaction. We would caution you about some (not all) mail-order firms — their prices may be heavily discounted but at a sacrifice of expert personal service. Some have limited guarantees and no after-sale service — if there is a problem you can find yourself on your own dealing directly with the manufacturer. Also, watch the shipping and packing charges!

ONE OF THE MOST POPULAR TELE-SCOPES for serious backyard astronomers is the 8-inch Schmidt-Cassegrain. It offers aperture, portability, and convenience. A wide range of models is available from Celestron International (such as their Ultima 8, at top) and from Meade Instruments (above). Above photo courtesy Meade Instruments.

EDITORS' CHOICES
UNDER $400 TELESCOPES

Model		Aperture	f/ratio	Mount	Slow-motion control	Motor type	Finder	No. of eyepieces	Focuser size (in.)	Retail price
REFLECTORS										
Edmund Scientific	Astroscan 2001	4.2	4.2	AA	–	–	–	1	1.25	$380
Meade Instruments	#4420	4.5	8	GE	√	opt.	5x24	1	1.25	$380
Orion Telescope	SpaceProbe 4.5	4.5	7.9	GE	√	opt.	6x30	2	0.965	$400
Coulter Optical	Odyssey 8	8	4.5	DOB	–	–	opt.	1	1.25	$275
Coulter Optical	Odyssey Compact	10	4.5	DOB	–	–	opt.	1	1.25	$350
REFRACTORS										
Orion Telescope	TeleVista	2.4	11.6	AA	√	–	5x24	2	0.965	$200
Jason Empire, Inc.	#307 Explorer	2.4	13.3	AA	√	–	5x24	3	0.965	$275
Meade Instruments	#291	2.4	15	GE	√	opt.	6x30	2	0.965	$280
Orion Telescope	Sky Explorer 60	2.4	15	GE	√	opt.	6x30	2	0.965	$350
Edmund Scientific	#31-611	2.4	8	EF	–	–	5x24	1	1.25	$390
Meade Instruments	#312	3.1	11	AA	√	–	5x24	1	1.25	$400

AA = Alt-azimuth; DOB = Dobsonian; GE = German equatorial; EF = Equatorial fork . Prices are actual retail as of mid-1991.

EDITORS' CHOICES
$400 to $1,000 TELESCOPES

Model		Aperture	f/ratio	Mount	Slow-motion control	Motor type	Finder	No. of eyepieces	Focuser size (in.)	Retail price
SCHMIDT-CASSEGRAINS										
Meade Instruments	#2045D	4	10	EF	√	DC	5x24	1	1.25	$800
REFLECTORS										
Tasco*	#11TR	4.5	7.9	GE	√	opt.	5x24	2	0.965	$450
*Same as Jason #327, Simmons #6450, and Swift #853.										
Celestron	C4.5	4.5	7.9	GE	√	opt.	5x24	1	1.25	$550
Parks Optical	f/6 or f/8 Precision	6	-	GE	opt.	opt.	6x30	1	1.25	$800
Parks Optical	f/6 or f/8 Astrolight	6	-	GE	√	opt.	8x50	–	2	$850
Celestron	SP-C6	6	5	GE	√	opt.	6x30	1	1.25	$900
Meade Instruments	#826C	8	6	GE	–	AC	8x50	1	2	$800
Meade Instruments	DS–10	10	4.5	GE	–	AC	–	1	2	$800
Coulter Optical	Odyssey 1	13.1	4.5	DOB	–	–	opt.	1	1.25	$575
REFRACTORS										
Meade Instruments	#323	3.1	11	GE	√	opt.	5x24	1	1.25	$460
Celestron	Firstscope 80	3.1	11.4	AA	√	–	6x30	1	1.25	$550
Orion	Sky Explorer 80	3.1	15	GE	√	opt.	6x30	2	0.965	$550
Tasco*	#17TR	3.1	11.3	GE	√	opt.	6x30	3	0.965	$600
*Similar to Jason #324.										
Celestron	SP–C80	3.1	11.4	GE	√	opt.	6x30	1	1.25	$750
Vixen/celestron	Custom 90M	3.6	11	AA	√	–	6x30	3	0.965	$650
Vixen/celestron	SP-90M	3.6	11	GE	√	opt.	6x30	3	0.965	$850

AA = Alt-azimuth; DOB = Dobsonian; GE = German equatorial; EF = Equatorial fork . Prices are actual retail as of mid-1991.

28 What about buying a used telescope?

If well cared for, a used telescope should perform as well as a new one. You can find telescopes in the classifieds in local newspapers and "bargain finders." You should also check with your local astronomy club or ASTRONOMY's Reader Exchange. A newsletter called *The Starry Messenger* (advertised in ASTRONOMY) is devoted to ads for used telescopes.

BUYING ACCESSORIES can keep you busy (and poor!) for years. But one or two extra eyepieces, a red flashlight, a good star atlas, and perhaps a few filters are all you really need.

29 What telescope would YOU buy?

This is impossible to answer. Someone who has been in the hobby for a while and who has already owned several telescopes would not select the same scope a first-time buyer would. Some people prefer the solidness and precision of a fine-quality refractor, others like the aperture and versatility of a Schmidt-Cassegrain, while others prize the light-gathering power and simplicity of a large Dobsonian reflector. There is no single best telescope. In fact, chances are the first telescope you buy will not be the last. Many backyard astronomers happily own two or three telescopes, each chosen for a certain type of viewing.

EDITORS' CHOICES
$1,000 to $3,000 TELESCOPES

Model		Aperture	f/ratio	Mount	Slow-motion control	Motor type	Finder	No. of eyepieces	Focuser size (in.)	Retail price
SCHMIDT-CASSEGRAINS										
Meade Instruments	#2080B	8	10	EF	√	AC	6x30	1	1.25	$1,150
Celestron	Classic 8	8	10	EF	√	AC	6x30	1	1.25	$1,200
Celestron	Powerstar 8–PEC	8	10	EF	√	DC	6x30	1	1.25	$1,800
Meade Instruments	Premier 2080-40	8	10	EF	√	DC	6x30	1	1.25	$1,850
Meade Instruments	Premier 2080-50	8	10	EF	√	DC	9x60	2	2	$2,200
Celestron	Ultima 8–PEC	8	10	EF	√	DC	8x50	2	1.25	$2,300
REFLECTORS										
Parks Optical	8" f/6 Precision	8	6	GE	opt.	opt.	6x30	1	1.25	$1,100
Parks Optical	8" f/6 Superior	8	6	GE	√	DC	8x50	2	2	$2,200
Parks Optical	10" f/5 Superior	10	5	GE	√	DC	8x50	2	2	$2,600
Tectron	15" Dobsonian	15	5	DOB	–	–	–	–	2	$2,250
Coulter Optical	Odyssey 2	17.5	4.5	DOB	–	–	opt.	1	1.25	$1,150
REFRACTORS										
Takahashi	FC–76ES Apo	3	7.9	GE	√	DC	7x50	1	2	$2,500
Vixen/Celestron	SP-Fluorite 90S	3.6	9	GE	√	opt.	6x30	2	1.25	$1,900
Celestron	SP-C102	4	9.8	GE	√	opt.	6x30	1	1.25	$1,250
Televue Optics	Genesis Apo	4	5	GE	√	opt.	opt.	1	2	$2,500
Celestron	SP-C102F Apo	4	8.8	GE	√	opt.	6x30	1	1.25	$2,500
Astro–physics	4" Starfire Apo	4	8	GE	√	DC	opt.	–	2.7	$2,950
Astro–physics	Star 12ED Apo	4.7	8.5	GE	√	DC	opt.	–	2.7	$3,000

AA = Alt-azimuth; DOB = Dobsonian; GE = German equatorial; EF = Equatorial fork . Prices are actual retail as of mid-1991.

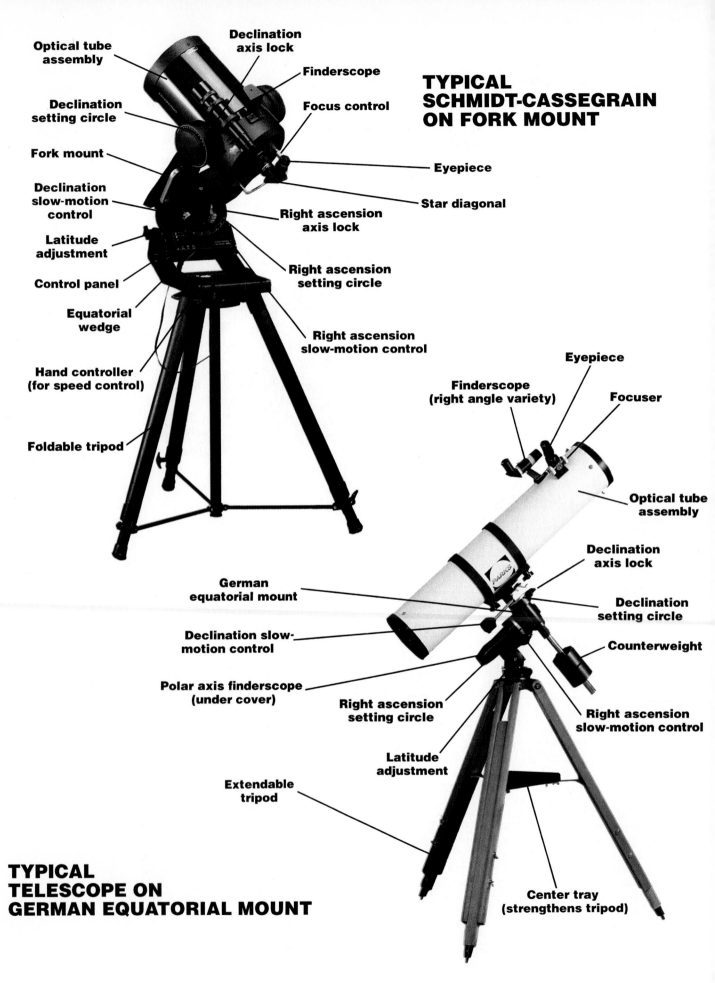

Optical tube
assembly

Declination
axis lock

Finderscope

**TYPICAL
SCHMIDT-CASSEGRAIN
ON FORK MOUNT**

Declination
setting circle

Focus control

Fork mount

Eyepiece

Declination
slow-motion
control

Star diagonal

Right ascension
axis lock

Latitude
adjustment

Control panel

Right ascension
setting circle

Equatorial
wedge

Right ascension
slow-motion control

Hand controller
(for speed control)

Eyepiece

Finderscope
(right angle variety)

Focuser

Foldable tripod

Optical tube
assembly

Declination
axis lock

German
equatorial mount

Declination
setting circle

Declination slow-
motion control

Counterweight

Polar axis finderscope
(under cover)

Right ascension
setting circle

Right ascension
slow-motion control

Latitude
adjustment

Extendable
tripod

**TYPICAL
TELESCOPE ON
GERMAN EQUATORIAL MOUNT**

Center tray
(strengthens tripod)

30 I have a child interested in astronomy. What scope should I buy? My budget is $200.

Avoid low-cost 500-power "department store" 50mm and 60mm refractors. Their poor mounts, eyepieces, and finderscopes will almost certainly make these telescopes a disappointment. The better 60mm refractors on alt-azimuth or equatorial mounts with slow-motion controls and a decent 6x30 finderscope *can* serve as starter scopes if your expectations are well-tempered. Acceptable models are available from astronomical dealers (such as those who advertise in ASTRONOMY) and local telescope stores.

But the truth of the matter is that for $200 (a common budget of parents with young astron-omers), there are few telescopes on the market we can endorse. Instead, we and many astronomy educators usually recommend a pair of 7x50 binoculars combined with a set of introductory books and star atlases, a package that will cost $100 to $200. Binoculars can reveal a surprising number of celestial objects (craters on the Moon, the moons of Jupiter, deep-sky objects such as star clusters and nebu-lae). A year spent exploring the sky with binocu-lars and a star chart can teach any novice astronomer, young or old, an immeasurable amount about the sky, the identity of stars and constellations, and the locations of celestial tar-gets. If your prospective astronomer is still inter-ested in the hobby after a year of binocular stargazing, then purchase a decent telescope for $400 to $500. At that point you will be more confident that your money will be well-spent. □

We trust we've been able to clarify the mysterious world of telescopes for you. To learn more about the various models on the market, we invite you to peruse the pages of ASTRONOMY magazine each month. In each issue you'll find advertisements telling you where to write to manufacturers and dealers for product literature and catalogs. You'll also find independent tests of products conducted by the staff of ASTRONOMY.

Please note that in this booklet we set an upper price limit of $3,000 for the most expensive telescope on our lists. However, you may be willing to spend more than that for a first purchase. If so, you have lots to select from, as the demands of ardent amateur astronomers have created a boom in high-end telescopes in recent years. Again, we refer you to ASTRONOMY for ads and information on these models.

Taking the time to research your purchase is a good idea. It's even a lot of fun. But don't agonize over the decision. With the exception of the department-store 60mm refractors we warned you about, there are few lemons on the telescope market. It's true that you may, like most backyard astronomers, soon outgrow your first telescope. But here's a final word of advice — long-time astron-omers who end up owning many telescopes over the years often find that their first telescope, as basic as it was, was the one that provided them the most enjoyment. The moral: keep your first telescope no matter what telescopes you later buy. You can then always recapture the thrill of your first sight of the universe.

Buyer's Checklist

√ **Does the scope have sufficient aperture?**
We suggest at least a 4-inch for deep-sky objects.

√ **How good are the optics?**
Will the dealer provide a guarantee of satisfaction?

√ **How steady is the telescope?**
After a light tap, vibrations should damp out in 1 to 2 seconds.

√ **How portable is the telescope?**
Can you carry it easily? Will it fit in your car?

√ **How easy is it to set up?**
Is the mount complicated? Heavy? Does it require tools?

√ **Does it have a drive motor? Is it AC or DC?**
DC drives can run directly from batteries and have a wider range of speed controls.

√ **Does the mount have slow-motion controls?**
These make it easier to aim the scope and follow objects.

√ **Does it have a separate finderscope?**
Finders that sight through the main optics are usually very poor.

√ **How large is the finderscope?**
A 25mm-aperture finder is poor, a 30mm OK, a 50mm best.

√ **What diameter eyepieces does it come with?**
0.965-inch-diameter models are usually poor. 1.25-inch is better.

√ **How good are the eyepieces?**
Orthoscopic and Plössl eyepieces are better than Kellners.

√ **Does it come with a case?**
It's useful if you will be transporting the scope.

√ **How expandable is the telescope?**
Is there a good array of accessories available?

√ **Does it come with a warranty?**
And who will honor the warranty with service?

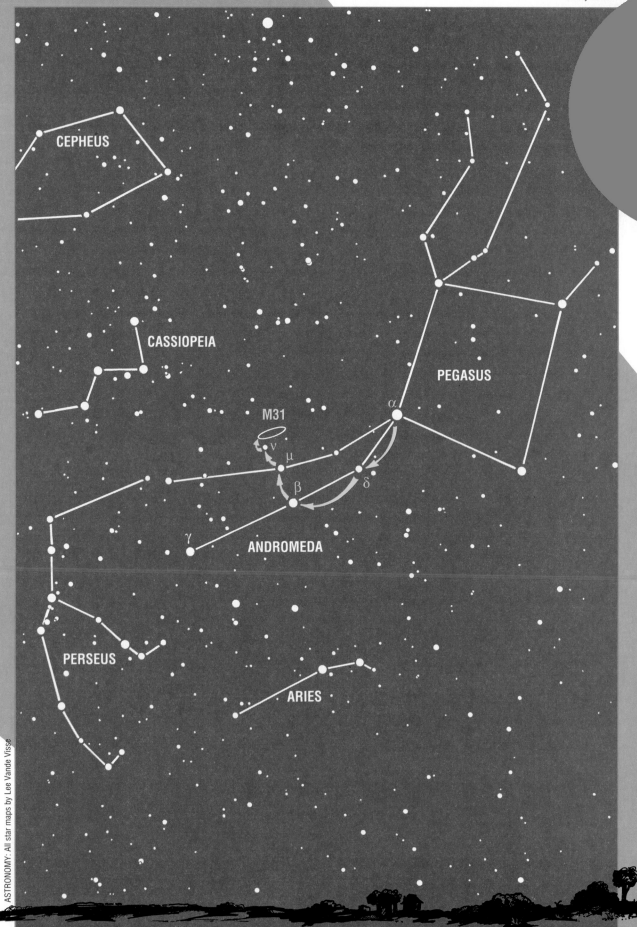

CEPHEUS

CASSIOPEIA

M31

ν

μ

β

δ

α

PEGASUS

γ

ANDROMEDA

PERSEUS

ARIES

Jeffrey Corder

ASTRONOMY: All star maps by Lee Vande Visse

OBSERVING TECHNIQUES

HOP INTO DEEP-SKY OBSERVING

Hopping from star to star can be the easiest and most rewarding way to locate deep-sky objects.

by Marcel W. Bergman

The Moon is New, the night sky steady and transparent. Starlight glints on the shiny tube of your new telescope. Multiple stars, star clusters, nebulae, and galaxies await you on this, your maiden voyage into deep-sky observing. How you prepare for this initial expedition can make all the difference in the world. You can enjoy hours at the eyepiece losing yourself to the splendors of deep space, or pore over star charts and peer through your finderscope only to find yourself deeply lost in space. You can finish the night with warm memories of your observations, or pack up early feeling cold and frustrated. Your first steps into deep-sky observing can profoundly affect the longevity of your new hobby. Too often,

tripping unprepared over the threshold of deep-sky observing culminates in the epitaph: "Telescope for sale. Rarely used."

For a successful observing session, you need to devise a strategy for finding your way. For a novice observer, locating faint celestial treasures in the inky voids that lie amid the chaotic array of stars is an imposing task. It certainly proved one to me. Before I bought my first telescope, I searched for advice on navigating the sky. I learned all I could about the two basic methods: setting circles and star hopping. In my experience, star hopping is the clear winner.

Moving Around in Circles

Setting circles are the dials you find on the two axes of an equatorial mount. The right ascension setting circle is the dial around the polar axis. The other setting circle surrounds the declination axis. To find a celestial object using setting circles, you first need to look up its right ascension and declination coordinates in an atlas, then dial in these coordinates on the setting circles, and *voila!* — the telescope should be pointing at the object. Unfortunately, in my experience, the scope is just as likely pointing at the ground.

There were other reasons that made me choose to star hop to deep-sky objects. Terrestrial navigation is ultimately based on the stars; surely the

stars are the most obvious beacons for journeys in space. My reading confirmed this logic, revealing phrases like, "M27, the Dumbbell Nebula, lies 3.3° due north of the star Gamma Sagittae." I never read instructions such as, "Find M27 at right ascension 19 hours 59.6 minutes and declination +22°43'." But what really made me choose star hopping was the advice of many veteran observers. They told me star hopping was the only way to really learn the sky. It made sense to me. I was bound to see and learn more on a trip that starts somewhere other than my destination.

Unfortunately, I never got specific advice on how to star hop. I figured it was easy. Just aim the scope along patterns of stars that progressively zero in on the target. It was a simple follow-the-dots. What more did I need to know?

Not So Fast

The first night with my brand new telescope dealt my star hopping career a major blow. For the first several minutes, I kept switching my attention between the finderscope view and the star chart, trying to lock onto a pattern in the myriad stars that filled the field of the finder. The inverted image I saw in the finderscope conflicted with my memory of the star pattern I had seen on the chart. Eventually, I forgot what it was I had set out to find.

I returned to my star atlas, red penlight in hand, to choose an easier object to find. Then, pasting my eye to the finderscope, I weaved the telescope this way and that, while the

ONE OF THE EASIEST STAR HOPS is to M31, the Andromeda Galaxy. M31 lies 2° northwest of Nu (ν) Andromedae, which itself is a simple four-step hop from 2nd-magnitude Alpha (α) Andromedae.

Van de Visse

SEPTEMBER 1992 **17**

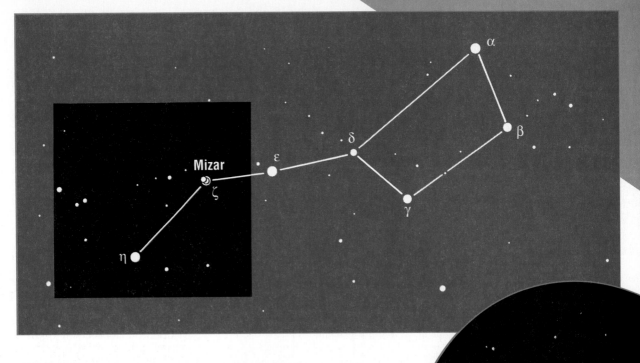

view in the finder wound that way and this. After several more minutes all I had to show for my efforts was a stiff neck and cold feet. I stumbled back to the star charts, my enthusiasm fading like the batteries in the red penlight that inadvertently had been illuminating the inside of my jacket pocket. I realized that celestial follow-the-dots was more complicated than the pencil-and-paper variety I grew up with.

Fortunately, star hopping improves with practice. I finally started finding some bright nebulae and star clusters. But whenever I looked for a faint galaxy or nebula, my hop slowed to a crawl. I decided to re-evaluate my whole star-hopping technique. Instead of plunging headlong into deep space as I had been, I now would hop methodically. I planned my next deep-sky observing session as though it would be my first.

Getting It Right

I found that to star hop efficiently, you need to familiarize yourself thoroughly with the set-up and operation of your telescope. If it's a new telescope, practice using it in daylight. Practice slewing the telescope on its axes. Practice using the slow-motion controls. The first time you use your telescope under the night sky, practice finding easy subjects like the Moon and planets using the same techniques you practiced in daylight.

As you build confidence in operating your telescope, also learn about your finderscope. If it has a right angle prism, it will produce an image that is right side up but reversed left to right. If it is a straight finderscope, the image will be both

THE BIG DIPPER SERVES as your starting point (above) for finding the spiral galaxy M101. From the double star Alcor and Mizar, hop east along a string of five stars until you reach the deep-sky landmark (right).

upside down and reversed left to right. Look through the finderscope while you move the telescope with the slow-motion controls. When you move the scope east or west, the view in the finderscope pans in the opposite sense. The same is true for movements north and south with a straight finderscope. Get these potentially disorienting effects straightened out in your mind. If you concentrate on the physical movement of the telescope, the view in the finderscope won't fool you.

The finderscope is your primary aid in star hopping. Collimate it accurately — in other words, make sure it points in precisely the same direction as the main telescope. It's easiest to collimate the finderscope in daylight. Point the telescope at a distant terrestrial object — a mountain, tall building, or other landmark a mile or more away works best — and center the object in the telescopic field of a low-power eyepiece. Then without moving the main telescope, adjust the finderscope so that the object is centered in the finder's field.

You'll also want to determine the size of your finderscope's field of view. Some finders have the field of view marked on their sides. If yours does not, take the scope out on a clear night and aim it at a pair of stars that just fit in the field. Then on a star

chart, measure the distance between the two stars. If the stars are 5° apart, then the circle of sky shown in your finderscope is 5° in diameter.

An alternative is to aim your telescope at a star near the celestial equator (a star with a declination close to 0°) and center the star in the finder's field. If you are using a clock drive, turn it off. Measure the time in minutes it takes for the star to drift out of the field and divide this time by two. The result is the angular diameter in degrees of the circle of sky that you see in your finderscope. For example, if it takes ten minutes for the star to drift from the center to the edge of the field, then your finderscope shows a patch of sky 5° across. Now when you want to move the telescope 5°, you know you need to pan through one field in the finderscope.

Once you've figured out the finderscope's field, you should make a

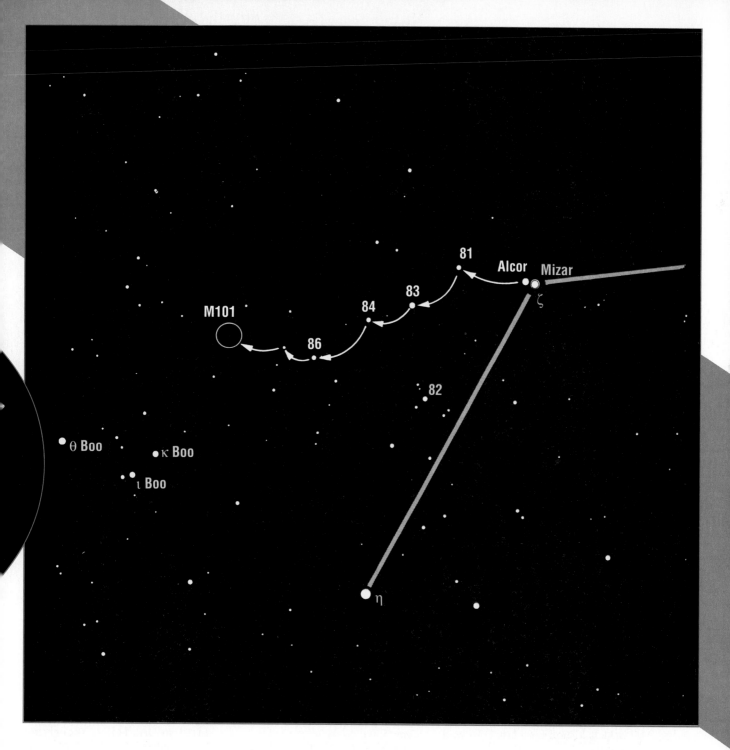

circular overlay of the same size for your star atlas. Simply determine the angular diameter of the finder's field at your atlas' scale and cut a circle of that size out of a piece of paper. You can then use the cutout to quickly compare what you see in the finder with the stars shown in your atlas.

Hop to It

Now you're ready for your first foray into star hopping. Luckily the autumn evening sky offers a nearly perfect deep-sky object for getting started — the Andromeda Galaxy, or M31. Not only is M31 bright enough to see with the naked eye from a dark-sky site (and from virtually anywhere with binoculars), but also all of the stars leading to it are visible without optical aid. You can star hop with confidence because if you lose your way, you can get back on track simply by glancing at the sky.

To find M31, pick a nice clear autumn evening when the Moon is either a slim crescent or out of the sky altogether. (In September, that means the first few days of the month or any time after mid-month.) After you arrive at your observing site, set up your telescope. While your eyes adapt to the darkness and your telescope adjusts to the ambient air tempera-

ture, locate the star at the beginning of your intended hop.

For the hop to M31, this is easy. Around 9 p.m. in mid-September, you can find the Great Square of Pegasus about halfway up in the eastern sky (see the map on page 60). Your starting point is the 2nd-magnitude star Alpheratz, also known as Alpha (α) Andromedae, which marks the northeast corner of the Great Square. Now center Alpheratz in your finderscope and confirm that its image is also in the center of the main telescope's field.

Your first hop takes you 7° east-northeast from Alpheratz to 3rd-

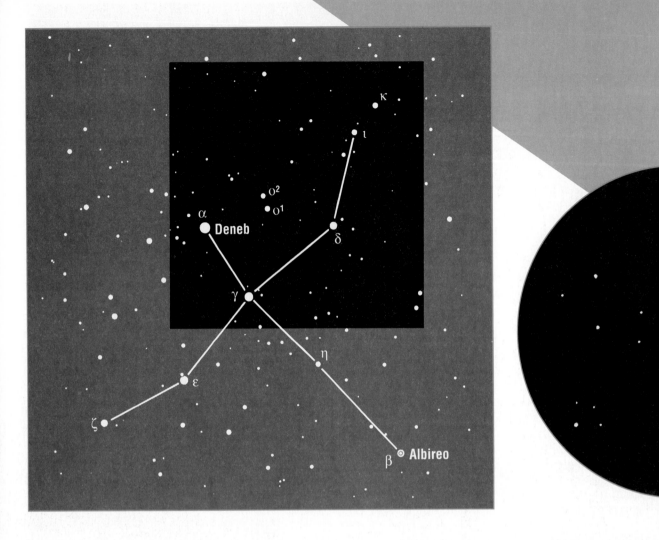

magnitude Delta (δ) Andromedae. Use the slow-motion controls to slew your telescope to this star, then sight along the scope's barrel to be sure you're pointing at the right star. Once you've found Delta, hop another 8° northeast to the 2nd-magnitude star Beta (β) Andromedae.

Next, hop 4° northwest from Beta to 4th-magnitude Mu (μ) Andromedae, then 3° north-northwest to 5th-magnitude Nu (ν) Andromedae. M31 is a short hop, just 1.5°, west of Nu.

Though star hopping to M31 is about as easy as it gets, it illustrates the techniques you'll need for other, more difficult star hops. First, make up a detailed plan for hopping to each of the deep-sky objects you want to observe. What you're looking for are recognizable star patterns that will lead you from a naked-eye star (preferably one 4th magnitude or brighter) to the deep-sky object. When you're out at your site waiting to get dark-adapted, reconnoiter the constellations you plan to explore and identify the naked-eye stars that will serve as your hopping-off points. When you are ready to hunt for a

deep-sky object, center the lead star and then use the slow-motion controls to star hop to your goal.

In my detailed plans, I include a description of the object's type, its location, and exactly how I intend to star hop to it. I also add brief comments on any interesting stars or objects along the way. For example, my entry for M31 looks like this:

Object: *Andromeda Galaxy — M31*
Type: *Spiral galaxy*
Constellation: *Andromeda*
R.A.: *0h42.7m*
Dec.: *+41°16'*
Star Hops: *Start at Alpheratz (Alpha Andromedae). Hop 7° east-northeast to Delta Andromedae, then 8° northeast to Beta Andromedae (double star). Next hop 4° northwest to Mu Andromedae and another 3° north-northwest to Nu Andromedae. M31 lies 1.5° due west of Nu.*

If you prefer, describe the star hops in terms of your finderscope's field instead of degrees. Either way, prepare the plan in advance so you don't waste precious time at your observing site fussing with star charts.

THE BLINKING PLANETARY (NGC 6826) lies in northeastern Cygnus (left), just a short hop from Gamma (γ) Cygni, the center of the Northern Cross (right).

A well-crafted, carefully designed plan will help make each trip to a deep-sky wonder a meandering route of memorable celestial scenery.

Another good target to test your star-hopping skills is the galaxy M101 in Ursa Major. Though this hop involves fainter stars than the one to M31 does, the pattern of the stars is easy to recognize. Start with the double star Alcor and Mizar in the handle of the Big Dipper, which lies in the northwest after darkness falls in September. (See the charts on pages 62 and 63.) From Alcor, hop 1.3° east to the star 81 Ursae Majoris (UMa). From there, follow a string of 5th- and 6th-magnitude stars to the southeast. First there's 83 UMa (1.2° away), then 84 UMa (0.9°), and then 86 UMa (1.3°). Then go 0.6° due east to the 7th-magnitude star at the northern end of a string of three stars. M101 lies 1.0° east-northeast of this star.

Our final star-hopping test is the Blinking Planetary Nebula, NGC 6826, in Cygnus, which lies nearly overhead on fall evenings (see charts above). The lead star in this star hop is Sadr (Gamma [γ] Cygni), the middle star in the Northern Cross asterism. From there hop 8° northwest to Delta Cygni and then 7° north-northwest to Iota (ι) Cygni. Next hop 2° south-southeast to the multiple-star system Theta (θ) Cygni, 1° due east to 6th-magnitude 16 Cygni, and finally 0.5° east to NGC 6826.

If you can't find your target after the final hop, try to spot it with averted vision; that is, look to the side of the object's suspected location. If you can't find the object in the finderscope, look through the eyepiece of your telescope and sweep the area slowly and systematically. If the object doesn't come into view after two or three tries, take a break and then move on to the next object on your plan.

By following a patiently planned observing strategy, I am learning the sky the way I always hoped I would. Every night the list of deep-sky marvels I can hop to with ease grows. I enjoy hunting for dim nebulae and galaxies, pausing to view an interesting multiple star or cluster along the way. And once I've hopped to an object a few times, I can find it quickly anytime in the future by just following the familiar pattern of stars.

Whether you're a novice observer standing on the threshold of deep space or a frustrated beginner on the verge of selling your telescope, don't despair. Prepare a deep-sky observing plan and on the next clear, moonless night, go ahead — hop to it.   ☐

Marcel W. Bergman is an avid amateur astronomer living in Calgary, Alberta.

Dialing for deep-sky objects

When star hopping doesn't work, use your scope's setting circles to find elusive objects.

by Mark J. Coco

Bruce Bond

"**D**arn! M76 should be right in the eyepiece. I know it's north of the bright star Almach, but I can't find it. There's got to be a way to find it."

There is: Use those mysterious circles on your mount. Hopping from bright star to bright star is an easy way to find some deep-sky objects. But when the object you want to observe doesn't have a convenient path of bright stars leading to it, you can use your scope's setting circles to guide you to the object.

All you need is a German equatorial or fork mount that is polar aligned and has setting circles. You don't need a clock drive, although that makes finding objects a little easier. Using either the direct indexing or offset methods described below, your circles can get you close enough to an object's position that a quick search of the region will reveal the object. To see how both of these methods work, let's first look at how astronomers map the sky.

Earth and Sky Coordinates

All stars appear fixed on a distant sphere called the celestial sphere. Earth is small compared to the size of the celestial sphere, so all observers on Earth are effectively located at the center of the sphere. To map the positions of stars and other objects on the celestial sphere, astronomers use several different coordinate systems. The one most useful for pointing telescopes is the equatorial system.

The equatorial system is based upon Earth's latitude and longitude system. Extending Earth's rotational poles out to the celestial sphere defines the north and south celestial poles. Earth's equator extended out to the celestial sphere defines the celestial equator, which divides the sky into two domes or hemispheres.

On Earth, the angular measure of a city from the equator is its latitude. Astronomers call the equivalent "celestial latitude" declination (abbreviated Dec). Ninety degrees of declination separate the celestial poles from the celestial equator. The celestial equator has a declination of 0°, declinations in the northern hemisphere have positive numbers, and declinations in the southern hemisphere have negative numbers. Because 1° covers a large amount of sky, astronomers divide the degree into 60 minutes (abbreviated ') and each minute into 60 seconds (abbreviated ").

Where the celestial pole and the celestial equator appear in the sky depends on your latitude. The point directly overhead, called the zenith,

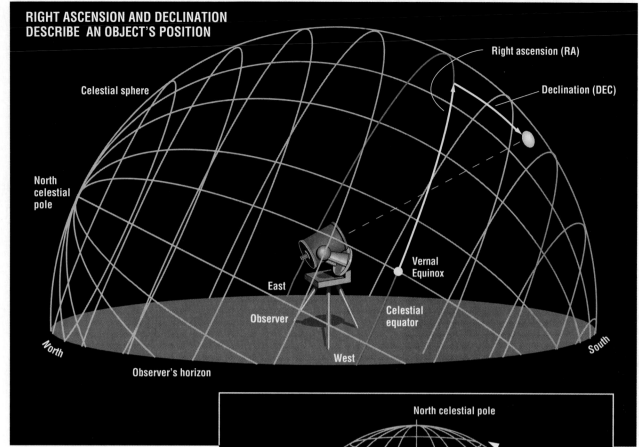

RIGHT ASCENSION AND DECLINATION DESCRIBE AN OBJECT'S POSITION

Right ascension (RA)

Declination (DEC)

Celestial sphere

North celestial pole

Vernal Equinox

East

Observer

West

Celestial equator

North

South

Observer's horizon

ASTRONOMY: all illustrations by Phil Kirchmeier

DECLINATION MEASURES an object's position above or below the celestial equator. Right ascension is the angular distance east from the vernal equinox.

appears at a declination equal to your latitude. The celestial equator lies an angular distance equal to your latitude from the zenith. The angular distance of the celestial pole above the horizon — the true horizon, not one cluttered with trees, hills, or buildings — is also the same as your latitude.

The second celestial coordinate, right ascension, mimics Earth's longitude. Tracing Earth's longitude lines out to the celestial sphere yields lines of right ascension (abbreviated RA). Because objects rise and set at different times, astronomers use hours to measure right ascension. Twenty-four hour lines mark right ascension on the celestial sphere, with each line marking one hour. One hour of right ascension covers a lot of sky (up to 15° at the equator), so astronomers divide each hour into 60 minutes (abbreviated m) and each minute into 60 seconds (abbreviated s). Don't confuse minutes and seconds of right ascension with minutes and seconds of declination. Four minutes of right ascension at the equator covers 1 degree of sky or the equivalent of 60 minutes of declination. At higher

EARTH'S ROTATION DEFINES THE CELESTIAL SPHERE

North celestial pole

Celestial sphere

Earth

Right ascension

Declination

Celestial equator

South celestial pole

declinations, the same four minutes of right ascension cover less sky because lines of right ascension lie closer together near the poles compared with near the equator.

While the celestial equator and poles provide a natural starting place to measure declination, right ascension can start anywhere. Earth longitudes have this same arbitrariness. By international convention, Earth longitudes start at Greenwich, England, with positive longitudes east of Greenwich and negative longitudes west. Right ascension 0h 0m 0s occurs at the vernal equinox, the point

EXTENDING EARTH'S POLES and equator out to the celestial sphere fixes the celestial poles and equator.

where the Sun crosses the celestial equator heading north on its apparent path annually through the sky. Astronomers measure other right ascensions eastward from this point up to a maximum of 23h 59m 59s.

Your telescope's equatorial or fork mount has two axes. One axis moves the scope in declination and the other, sometimes called the polar axis, moves it in right ascension.

FEBRUARY 1993 **23**

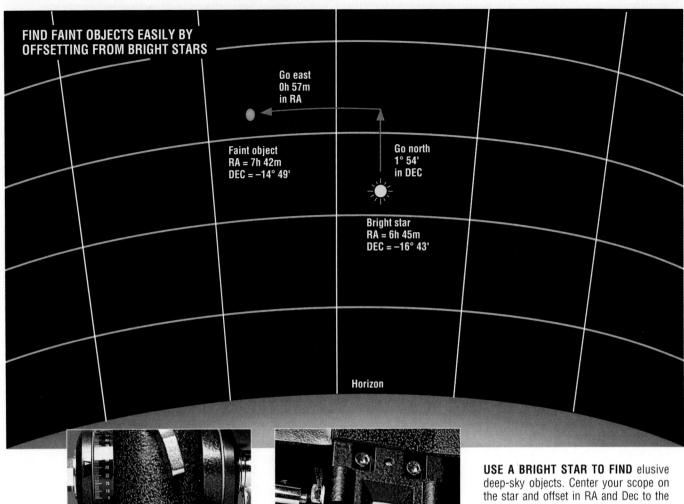

Go east
0h 57m
in RA

Faint object
RA = 7h 42m
DEC = −14° 49'

Go north
1° 54'
in DEC

Bright star
RA = 6h 45m
DEC = −16° 43'

Horizon

ASTRONOMY: both photos by Darla Gawelski

RIGHT ASCENSION INCREASES to the east. This circle, which has marks every 10 minutes, reads 6h 45m.

DECLINATION INCREASES to the north. This circle has divisions every 2° and reads −17°.

USE A BRIGHT STAR TO FIND elusive deep-sky objects. Center your scope on the star and offset in RA and Dec to the faint object.

Each axis has a setting circle. The Dec circle has divisions every one or two degrees and ranges from −90° to +90°. The RA circle has 24 divisions marking every hour and additional divisions marking every 5 or 10 minutes.

If the RA circle on your scope has divisions reading 0 to 6 and back to 0, you have an hour angle circle instead of a RA circle. A circle divided into 24 hours can also read hour angle and not right ascension. If you move your scope to the east, the number on the circle should increase. If it doesn't, you have an hour angle circle. Don't

worry. You can use an hour angle circle to find deep-sky objects.

Using Your Setting Circles

Backyard astronomers use setting circles in two ways. In offsetting, you use the circles to move from a bright star to the object of interest. You can use this technique with any scope that has setting circles. With direct indexing, you use the circles to dial in an object's coordinates directly. In order for you to use this technique, your scope must have an RA circle and not an hour angle circle.

Both of these techniques require

that you polar align your scope. Good polar alignment is needed for offsetting your scope; precise polar alignment is needed for direct use of the circles. (See "Polar Aligning Your Telescope" in the May 1992 ASTRONOMY). If you plan to use offsets to find deep-sky objects, skip to the section entitled "Offsetting to Objects." If you plan to use direct indexing, you need to calibrate your setting circles.

Polar alignment automatically calibrates the Dec setting circle. The reason for this is that the lines of declination remain constant in the sky. So once you have the scope aligned to the pole, the circle should always be calibrated. If the circle has slipped, it is easy to recalibrate. With your scope polar aligned and the scope pointed at the pole, the Dec circle should read 90° degrees — +90° for observers in Earth's Northern Hemisphere and −90° for those in the Southern Hemisphere. If it doesn't read 90°, loosen the screw holding the circle to the declination axis and rotate the circle until it reads 90°. If your Dec circle doesn't have a screw but is glued to the axis instead, bend the indicator slightly so the circle

Reading Vernier Scales

Vernier scales enable you to read setting circles more accurately. Typically you can determine an object's position to within 1 minute in RA and 15 arcminutes in Dec using verniers.

The zero mark on the RA vernier is the RA indicator. If the mark lines up exactly with one on the circle, then the RA is exactly the number on the circle. Commonly the indicator points between two marks on the circle. In the example at right (the vernier is compressed for clarity), the indicator points between 6h 40m and 6h 50m. So the base RA is 6h 40m. Determine the exact number of minutes by finding which vernier mark lines up with a mark on the circle and add this number to the base RA. In the example, the 5-minute mark lines up with 7h 20m. Add 5 minutes to 6h 40m to get 6h 45m.

Dec circles have two verniers, one for going south and one for going north. If the zero indicator lines up with a mark on the Dec circle, then the Dec is exactly the number on the circle. In the example, the indicator lies between –16° and –18°. Going north, the southernmost number (–18°) is your base Dec. Using the right vernier, the 1° 15' mark lines up with 8° on the circle. You add this number to the base Dec. So in the example, –18° 00' plus 1° 15' yields a Dec of –16° 45'.

Going south, use the left vernier to obtain the Dec. In the example, the 45' mark lines up with –22°. Subtract this number from the northernmost Dec near the zero indicator. So in the example, –16° minus 45' yields a Dec of –16° 45'.

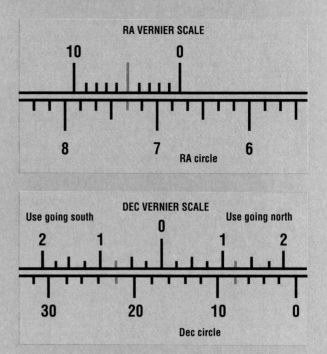

reads 90°, or break the glue bond, carefully set the circle, and reglue it.

If you plan to use direct indexing and you have an RA circle and a clock drive, check to see if your circle is driven. Point your scope east, note the right ascension marked by the indicator, and turn on the clock drive. Come back in a half-hour or so and read the RA circle again. If the circle reads the same as it did initially, your circle is driven. If the circle reads a lower right ascension, then your circle is not driven. Most Schmidt-Cassegrain telescopes have driven circles, but many German equatorial mounts do not.

For driven circles, calibrating the RA circle need be done only once at the beginning of the night. For non-driven circles, you need to calibrate before you search for each object. Calibration of the RA circle is necessary because the stars and the vernal equinox move across the sky due to Earth's rotation and orbital motion around the Sun. You can keep track of the motion of the vernal equinox using a special time system called sidereal time, but a simpler way exists.

Find a bright star in your scope and, if you have one, turn on the scope's clock drive. Look up the star's right ascension in a star catalog and then simply turn the RA circle

until the indicator points to this number. Your RA circle is now calibrated.

Direct Indexing

Before you start observing, prepare a list of objects you wish to find, including each object's coordinates and the coordinates of one or more bright stars for RA circle calibration. Don't use a star chart to obtain the objects' coordinates; it provides only an approximate set of coordinates. Use a catalog or other reference book to look up the coordinates.

Armed with the correct coordinates, go to the telescope and calibrate the RA circle. Then move your scope until the circles indicate the coordinates of the object you wish to find. An easy object to find is the Beehive cluster (M44) in Cancer at right ascension 8h 40m and declination +20° 00'. The object should appear in the finder.

To move to the next object, simply move the scope to the object's coordinates if you have a driven RA circle. If not, you don't have to recalibrate the RA circle using a bright star. Simply rotate the RA circle to the present object's right ascension before you move the scope to the next object.

In a perfect world, moving the telescope until the circles indicate the proper coordinates should put the object right in the center of the field of

your telescope. But this doesn't always occur. Several factors influence the accuracy of your setting circles. One is polar alignment; inaccurate alignment throws off the reading of your setting circles. While setting up, take a little extra time to align your polar axis as precisely as possible.

A second factor is the epoch of coordinates you use for calibrating the RA circle and the coordinates of the objects sought. Precession, a slight wobble of Earth's rotational axis, causes the positions of stars to shift. Over a period of 50 years, precession can cause an object to shift by as much as 16' in declination and 3 to 15 minutes in right ascension. So astronomers compile catalogs of star and object positions for different dates or epochs. Most catalogs now printed use epoch 2000.0 coordinates, but older catalogs may use 1950.0 or even 1900.0 coordinates. Which epoch you use doesn't generally matter as long as you use coordinates from the same epoch for both RA calibration and the target object.

Using 1950.0 coordinates for a bright star to calibrate the RA circle, but 2000.0 coordinates for a deep-sky object can cause you to miss the object. If you need to use different epochs, you can find tables — and equations, if you prefer them — to convert coordinates from one epoch

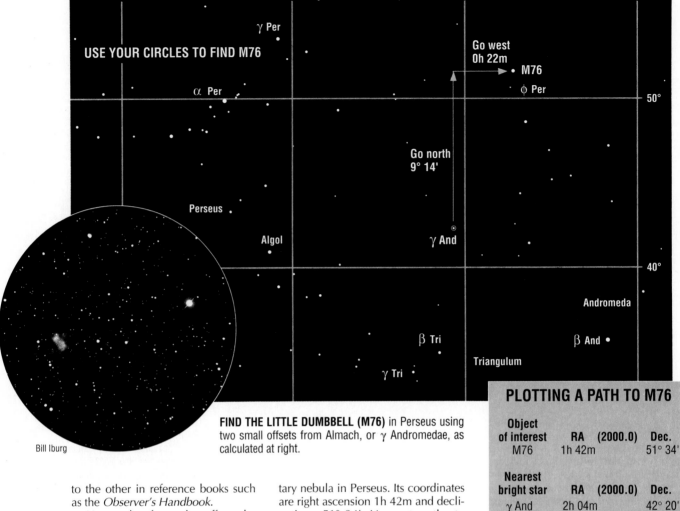

USE YOUR CIRCLES TO FIND M76

4h · 3h · 2h · 1h

γ Per

Go west
0h 22m
→ · M76
φ Per

α Per · 50°

Go north
9° 14'

Perseus

Algol · γ And

40°

Andromeda

β Tri · β And ·

Triangulum

γ Tri

Bill Iburg

FIND THE LITTLE DUMBBELL (M76) in Perseus using two small offsets from Almach, or γ Andromedae, as calculated at right.

to the other in reference books such as the *Observer's Handbook.*

Yet another factor that affects the accuracy of your setting circle readings is the circles themselves. Most Dec circles have a marker every one or two degrees. RA circles typically have divisions every 10 minutes. Finer divisions are useful for finding faint objects not easily seen in a finderscope. Some manufacturers provide vernier scales, a finely divided scale that augments the setting circles, to enable you to obtain more accurate readings. See Reading Vernier Scales on the previous page.

Offsetting to Objects

Setting circles can still help you find faint deep-sky objects even if your scope has only an hour angle circle or if you don't want to bother with a precise polar alignment. A clock drive makes offsetting easier but it's not necessary.

Instead of calibrating the RA circle as you must do for direct indexing, you use differences in right ascension and declination between the object and a bright, nearby star. Suppose you want to find the Little Dumbbell (M76), a faint 11th-magnitude plane-

tary nebula in Perseus. Its coordinates are right ascension 1h 42m and declination +51° 34'. Use a star atlas to find a bright star nearby. (Remember to get its position from a catalog, not from the star atlas.) The star should be easily visible in your finderscope and close to the object so errors in the polar alignment don't affect the offset. Also, if you don't have a clock drive, a nearby star prevents the object from drifting much while you move the scope from the star to the object.

Many bright stars lie near M76. The closest is γ Andromedae, a 2.3-magnitude star that is also called Almach. It lies at right ascension 2h 04m and declination +42° 20'. The difference in Almach's and M76's positions is 0h 22m in right ascension and 9° 14' in declination. You should compute these offsets before you set your scope to Almach.

Find Almach and center it in your scope. Note the readings of the setting circles. Move the scope west in right ascension 22 minutes and then north 9° 14'. You may need to round these numbers slightly, depending upon the divisions on your circles, say to 20 minutes in RA and 9° 15' in Dec. This object may not be visible in

your finder, so sweep the region carefully with a low-power eyepiece in your telescope.

Whether you use offsets or direct indexing, setting circles can help you find objects too faint to be seen in you finderscope or objects that don't have good star-hopping paths. Learning to use your circles can open a new realm of objects for you to observe. And while you don't need to use setting circles to enjoy the night sky, they can make it easier to find some elusive objects. □

Mark J. Coco is a technical writer and amateur astronomer living in Southern California.

PLOTTING A PATH TO M76

Object of interest	RA (2000.0)	Dec.
M76	1h 42m	51° 34'

Nearest bright star	RA (2000.0)	Dec.
γ And	2h 04m	42° 20'

Subtract bright star position from object's position to obtain offsets.

Offsets	–0h 22m	9° 14'
Direction	go west	go north

Magellan Scores at Venus

A cloud-penetrating spacecraft strips Venus of its worn-out image and reveals a world active with volcanoes, mountains, surface cracks, and craters — an environment unique in the solar system.

by David J. Eicher

THE MAGELLAN SPACECRAFT arrived in orbit around Venus on August 10, 1990, promising to give astronomers their best look at Earth's "sister planet." After a few initial communications glitches, the craft performed admirably and returned radar images that astonished planetary scientists. At right three deep impact craters ranging from 23 to 30 miles across lie on a heavily fractured plain. Many volcanic domes are visible at lower right. Streaks north of the largest crater may represent sand dunes. All images courtesy NASA/JPL.

"This is absolutely some of the finest astronomical imaging ever collected!" As he scans several dozen pictures of Venus, Project Scientist R. Stephen Saunders has good reason for excitement. On August 10, NASA's *Magellan* spacecraft entered orbit around Venus after a 15-month, 950-million-mile-long journey. After several heart-stopping communications blackouts, the spacecraft began its mission of mapping the planet by radar on September 15. The mission is scheduled to last 243 days. *Magellan*'s images depict Venus in unprecedented detail, revealing objects as small as 400 feet across. (The previous best images, sent back by the Soviet *Venera 15/16* craft, had a maximum resolution of just under one mile — only 1/10 as good as *Magellan*'s.)

Although NASA has released only a trickle of *Magellan*'s first images, the quality of the new images sent a thrill of jubilation through the agency's *Magellan* staff. "It will be years before the images are analyzed," remarks Saunders. "But we can already see clearly a planet that has only been seen hazily before *Magellan*." James Head, a mission geologist borrowed from Brown University, believes the spacecraft is "giving us a revolutionary new view of Venus."

What's in the new views of Venus? "We've just started to examine them," says Saunders, "and the spacecraft has so far covered a very small area on the planet. We've seen all the familiar features we expected to see and some relationships that are really beginning to intrigue us."

"The most prominent things we expected to see are craters," declares Saunders. "We've produced a mosaic of a field in the Lavinia region that we call the 'crater farm' because of its three prominent craters and radar-bright blankets of ejecta. At first glance, Venusian craters appear similar to other solar system craters, like those on the Moon. We're seeing all of the typical characteristics — strong central peaks, terraced walls, and volcanically flooded floors. However, Venusian craters differ from lunar craters in that they have no bright rays caused by splatter — which does not happen easily in the thick Venusian atmosphere — and they have a radar-dark shock ring around

VENUSIAN CRATERS show terraced inner walls and a strong central peak, strikingly similar to craters on the Moon, Earth, and Mars. *Magellan*'s radar image of the crater Golubkina (below) shows details as small as 400 feet across, a tenfold increase in resolution over the previous best image collected by the Soviet *Venera 15/16* spacecraft (left side of the image). A three-dimensional computer reconstruction of *Magellan*'s picture of Golubkina reveals the third dimension of the structural features of the 20-mile-diameter crater (right).

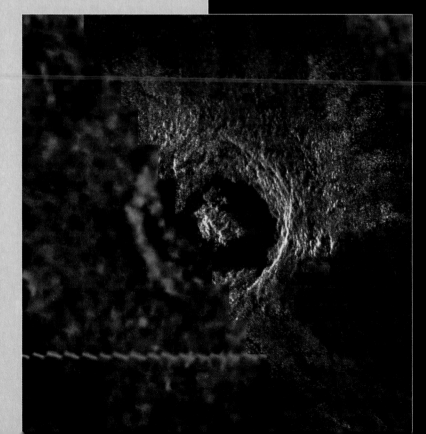

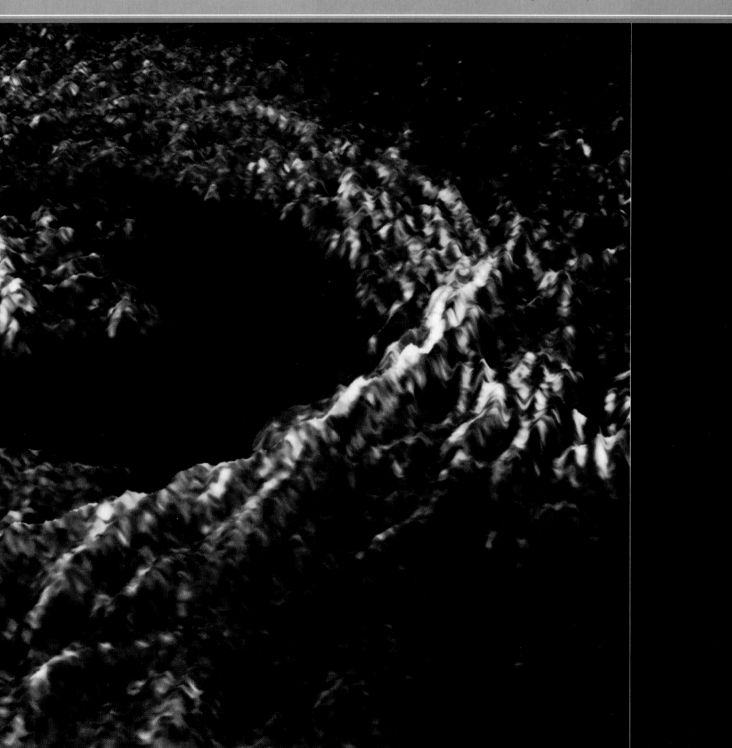

At first glance, Venusian craters look much the same as lunar craters. Yet Venus' craters are large and lack rays because of the planet's heavy atmosphere.

ring in this image centered on a crater less than one mile across. Surface deposits appear bright in radar because they are rough.

overrun by a rough lava bed that appears bright and extends from the center to the upper left corner. This image covers an area measuring about 20 by 47 miles.

Is Cleopatra Terra an impact or volcanic feature? Does Maxwell Montes have huge faults like San Andreas? What are the secrets in these piles of rock?

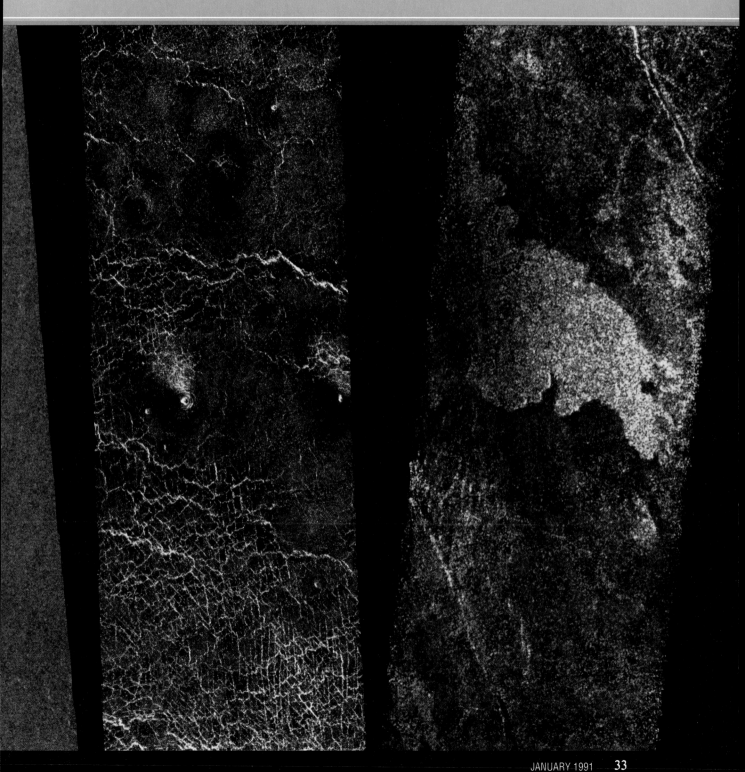

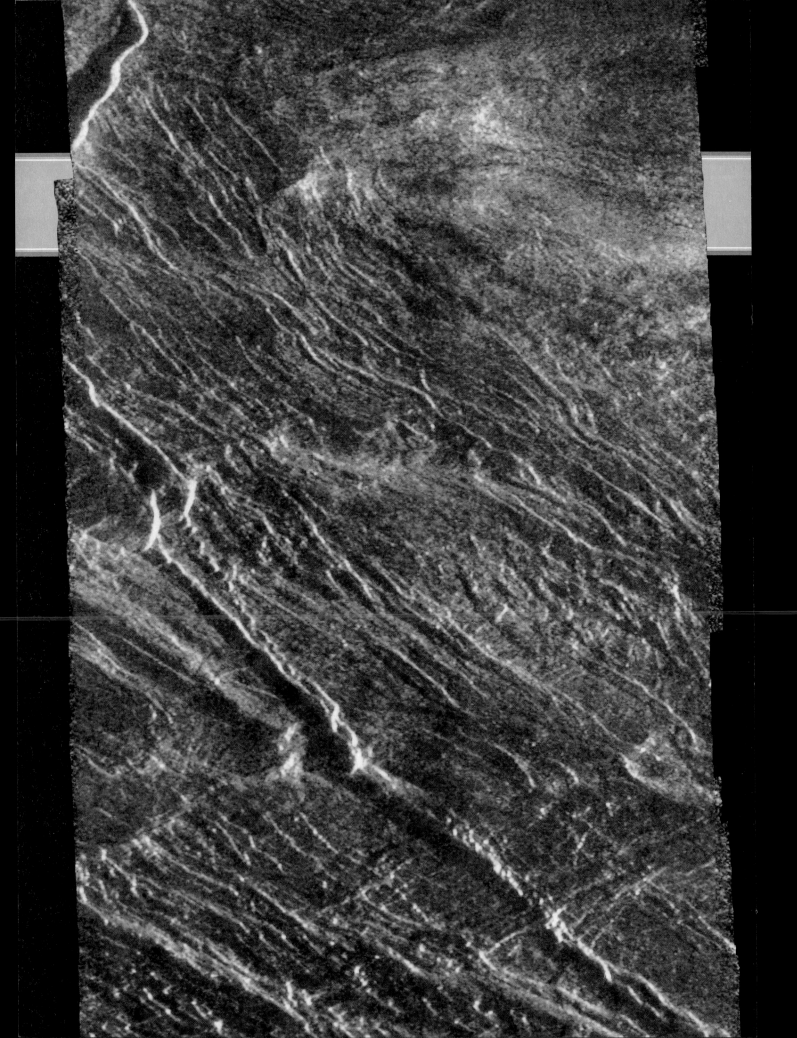

MOUNTAINS AND VALLEYS run side by side for dozens of miles in the Lakshmi region of Venus. The twisting mountains rise nearly two miles above the floor of the plain, Lakshmi Planum, below.

Venus' surface appears to be old, as if great activity happened early, resurfacing parts of the planet, and not much has happened since.

the ejecta. This is caused by a shock front created in the planet's atmosphere. Also, there are no small craters on Venus. Only large hunks of material make it through the atmosphere to the planet's surface."

Saunders points out another curious feature of the Venusian craters. "Perhaps the most startling thing about the craters is they all look the same," he says. "They're all fresh looking. In some areas there doesn't seem to be much going on that changes the planet's surface over time. This is amazing to me because it means that we might be looking at regions of a planetary surface that have been relatively unchanged from something like a billion years ago. This will be one of the big questions as *Magellan* research continues: Is Venus a planet that wants to turn itself inside out by volcanism every couple hundred million years and then just sit there for a while?"

ROLLING HILLS AND FRACTURED TERRAIN dominate the Phoebe region. The bright features in the top half of this image are hills; the dark terrain below is a low elevation plain, probably formed by overlapping volcanic flows. The conspicuous fault at center is just 500 yards wide.

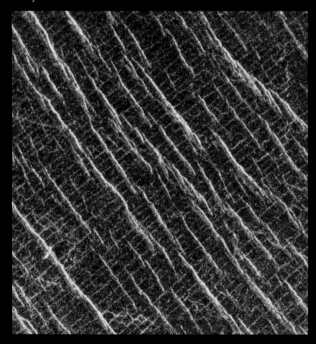

VOLCANIC CALDERAS mark the low-elevation plains in the Phoebe region. The caldera walls appear bright most likely because they are rough; dark features mark lower elevations. The larger of the two measures 5 by 2 miles in extent.

RADAR-BRIGHT PARALLEL LINES serve up a strange puzzle for astronomers, who believe the lines are either fracture or fault features. These features have not been previously observed on Venus or elsewhere in the solar system.

Magellan represents some of the most important astronomical images ever obtained — scientists will be sorting through them for years to come.

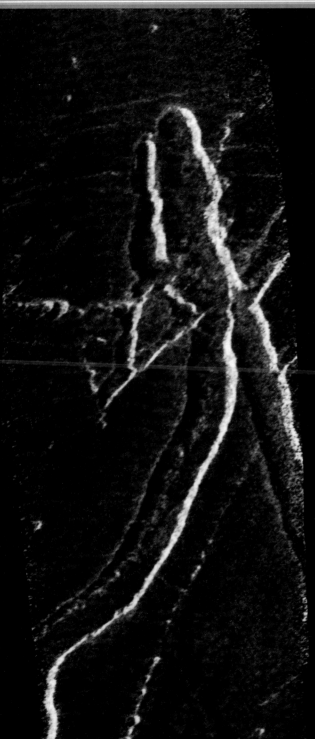

The planet's craters aren't its only geological attraction. Images sent back by *Magellan* show a detailed series of fracture and fault lines, volcanic features, and mountain chains that extend for hundreds of miles. "For example, in one image mosaic we call the crater farm," remarks Saunders, "we have detected a radar-bright patch of material extending out of the northern rim of the largest crater. If you look at where the bright patch north of the crater brightens and disappears, you can see some little glints that are suspiciously linear. That's just the way sand dunes look in radar when you get reflections off the steeper slip faces. We may have discovered the sand dunes of Venus!"

A volcanic dome centered in one of the images generated much excitement among the scientists. "Venus contains lots of little volcanic features here and there just like those on Earth," according to Saunders. "They're just right — the correct size, shape, and morphology. But I haven't yet seen any volcanic features that appear to be recent, and this is very surprising. I'm sure we'll eventually find recent volcanic activity, but so far — even in a volcanic plain — we've seen none." However, *Magellan* has thus far explored only a small percentage of the planet's surface.

"The really exciting part of the *Magellan* mission is the part to come," says Saunders. "We'll see images of Maxwell Montes, a 7-mile-high mountain that is the highest point on Venus. We'll look at Cleopatra Terra, a 60-mile-diameter feature that may have either a volcanic or impact origin. Examining these and other areas will allow us to answer many questions we've had about Venus for years, and perhaps tighten our vision of how the planet works as a system. After all of the delays and the occasional glitch, it's almost unbelievable to think that the mission is now happening. People generally don't realize how important this mission is to our understanding of the solar system overall and to Earth science as well. We'll hear reverberations from this mission for years to come."   □

DEEP TROUGHS ENCASED BY FAULTS characterize the most unusual feature in the planet's Lavinia region. The origin of this 50-mile-long feature is not clear. Unraveling the details of this and other features on our sister planet will challenge *Magellan* scientists for years to come.

An Observer's Guide to Sunspots

Did you know that the Sun is the only star you can observe in detail and that observing it doesn't even require dark skies?

by Edmund Fortier

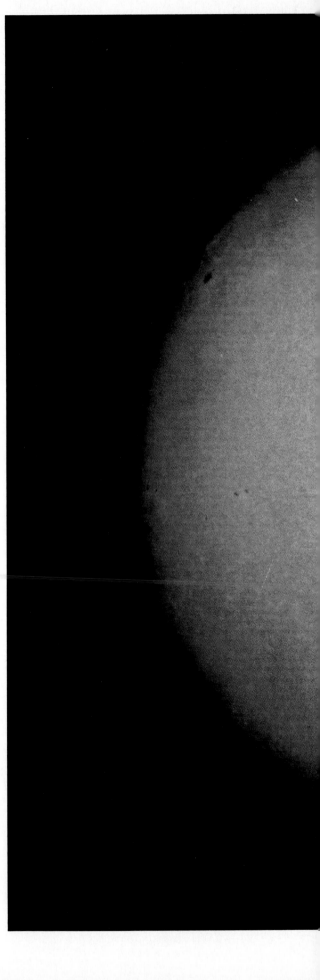

Every eleven years, as a new solar cycle builds toward maximum, astronomers witness a tremendous surge in sunspot activity. At such times these magnetic storms can, in the course of a few days or weeks, evolve into huge, finely detailed structures. In fact, it is not unusual for some sunspots to measure 25,000 miles or more across — large enough to be seen with the naked eye.

If you have been skywatching for less than ten years, you've never observed the Sun during its most active phase. The last peak in the solar cycle occurred in 1980. By mid-decade, activity had declined to minimum and sunspots had all but vanished from the surface of the Sun. Now the cycle is again at its eleven-year peak. Because the 1990 maximum was exceptionally active — and we are still in a period of high sunspot activity — this is an ideal time to take a close look at the evolution of a typical sunspot group and to consider the various features a careful observer can expect to see.

First, though, a word of caution: Never look at the Sun unless your telescope is properly fitted for solar observing. For sunspots, use a white-light filter. These filters, usually made from aluminum-coated Mylar film or Inconel-coated glass, attach to the front end of the telescope or can even be held in front of your entire face when looking at the Sun with the naked eye. Mylar filters produce a cool, bluish image of the Sun, which some observers find objectionable. Inconel filters, on the other hand, render the solar disk a warm and natural yellow-white, but they are more expensive. Whichever filter you choose, be sure that it fits securely and don't forget to cover your finderscope.

Now that you've heard the safety lecture, let's examine exactly what sunspots are, anyway. They are phenomena of the solar photosphere, the region of the Sun that emits most of the radiation we see. When you're next observing the Sun with a telescope, look closely at the center of the solar disk, and if the seeing is good, you'll detect a subtle "rice grain" texture called granulation. The photosphere is made up of several million individual granules. These tiny structures, which are actually a few hundred miles across,

SUNSPOTS, EASY AND FUN TO OBSERVE,
offer changing patterns as the Sun rotates
day by day. Photo by Lee C. Coombs.

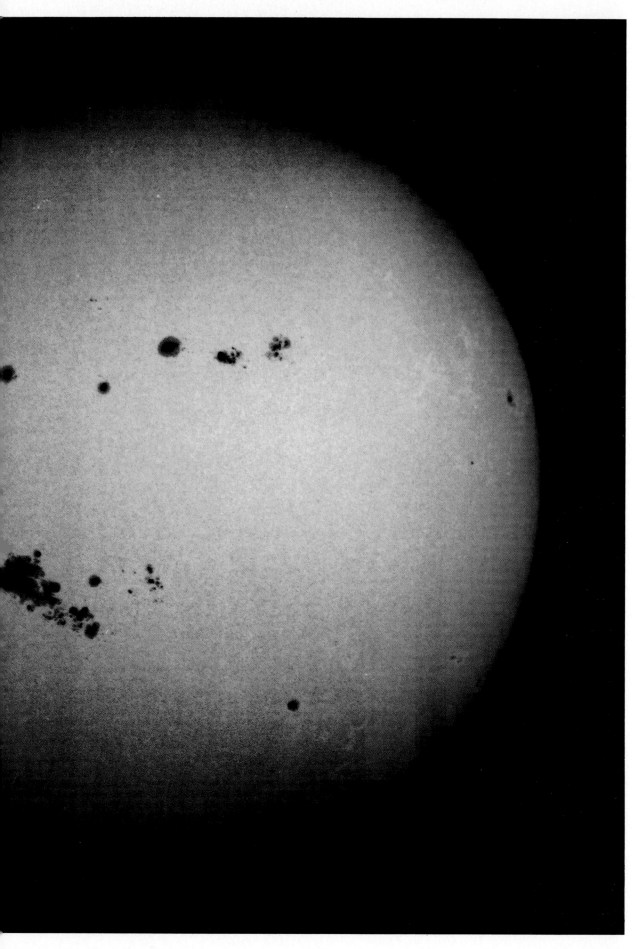

GIANT MAGNETIC STORMS, sunspots show dramatic details in small telescopes — multiple groupings, dark umbral centers, lighter penumbral areas, and whitish faculae.

have an average lifetime of about ten minutes and are caused by convection, the rising and sinking of cells of hot gas. Granulation provides us with direct visual evidence of a convection cycle in the Sun.

You will quickly discover, however, that granules are not visible across the entire solar disk. The photosphere appears noticeably darker toward the edge of the Sun, and our view there is obscured. This phenomenon, known as limb darkening, occurs because light from the limb must pass obliquely through a greater thickness of photosphere, limiting our view to the cooler upper layers. In contrast, when we look at the center of the solar disk, we see deeper into the hotter and brighter layers of the Sun's atmosphere. Because of the contrast provided by the limb-darkening effect, the periphery of the Sun is the only place you can see the bright, mottled patches known as faculae. But faculae exist everywhere on the Sun, and astronomers think they are precursors to sunspots. These whitish streaks always precede the development of a sunspot, and they always survive its final dissolution.

How does a sunspot develop? All spots begin

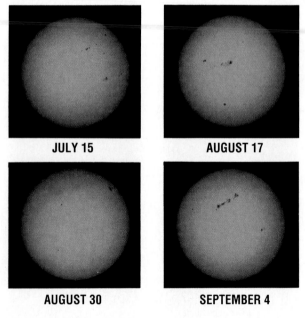

JULY 15   AUGUST 17

AUGUST 30   SEPTEMBER 4

SIX WEEKS ON THE SUN shows the changing appearance of spots. Rick Dilsizian's sequence shows the Sun on July 15, August 17, August 30, and September 4, 1989.

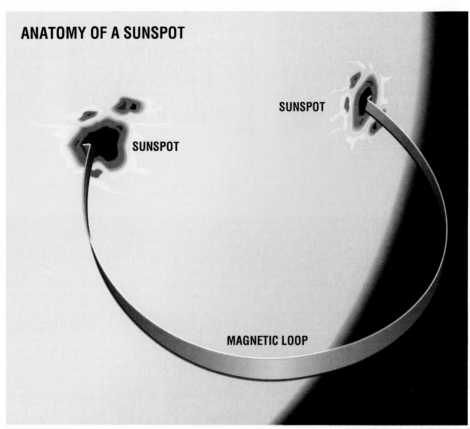

ANATOMY OF A SUNSPOT

SUNSPOT

SUNSPOT

MAGNETIC LOOP

SUNSPOTS ARE AREAS on the Sun's surface pierced by magnetic loops from inside the Sun.

MAGNIFICENT DETAIL is visible in sunspot groups, like this one photographed by Gregory Terrance.

EVOLUTION OF A SUNSPOT

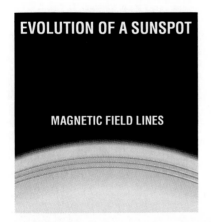

MAGNETIC FIELD LINES

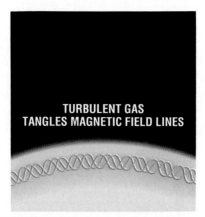

**TURBULENT GAS
TANGLES MAGNETIC FIELD LINES**

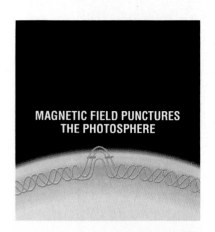

**MAGNETIC FIELD PUNCTURES
THE PHOTOSPHERE**

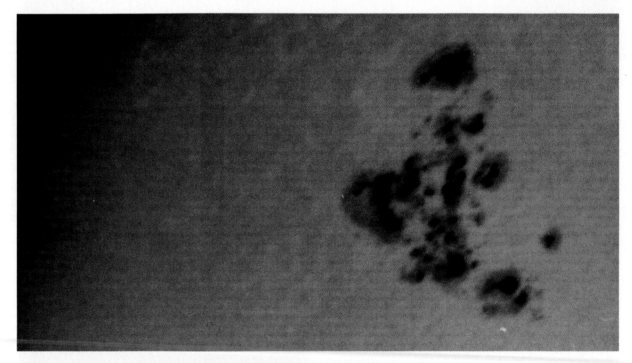

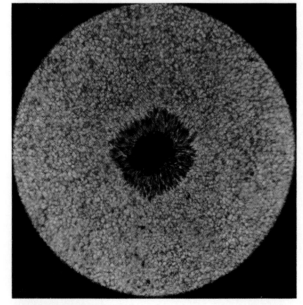

A SINGLE SUNSPOT seen close up shows magnificent detail. Sacramento Peak Observatory photo.

LARGE SUNSPOT GROUPS contain dark umbral centers, gray penumbral haloes, many large and small single and overlapping spots, and surrounding whitish plages. Photo by John S. Hicks.

when a normally bright granule suddenly darkens, becoming a tiny, barely discernible spot called a pore. Most pores quickly disappear, but a few grow in size, developing a dark core called the umbra and the beginnings of a lighter, filamentary structure called the penumbra. Most spots dissolve at this point, but a few continue to grow, and within a day or two evolve into mature spots with large umbrae and full, symmetrical penumbrae. The temperature of such a sunspot falls to about 4000 K, some 2000 K cooler than the surrounding photosphere.

Although sunspots may develop as single entities, they can also develop in magnetically associated groups. These sunspot clusters are usually bipolar; that is, each contains sunspots of both polarities. In considering the evolution of these sunspot groups, it's important to keep two things in mind. First, a spot group may dissolve at almost any stage in its development,

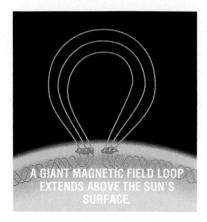

A GIANT MAGNETIC FIELD LOOP EXTENDS ABOVE THE SUN'S SURFACE.

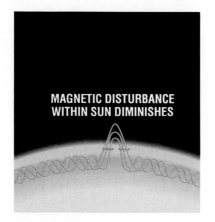

MAGNETIC DISTURBANCE WITHIN SUN DIMINISHES

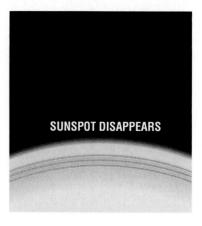

SUNSPOT DISAPPEARS

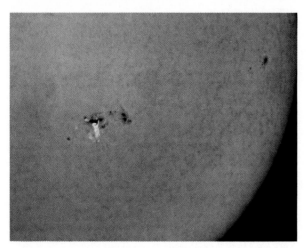

A GIANT FLARE occasionally occurs in association with a sunspot. This hydrogen-alpha photo depicts an event that may also have been visible in white light. Courtesy NASA/ NOAA Solar Observatory.

and only a handful pass through every phase of the cycle. Second, since the Sun rotates once every 27 days, two weeks is the longest you can keep a single group under constant surveillance.

Even the most complex sunspot group begins inauspiciously, with the sudden appearance of a few pores on the solar disk. Within a day the incipient group appears as two concentrations of umbral spots, elongated in a roughly east-west direction. One spot in each concentration usually develops more fully than the others.

During the first few days little of telescopic interest occurs. Around the second week of its existence, a sunspot group reaches maturity. The distance between the two main spots, which has been steadily increasing, reaches a maximum separation on about the tenth day. An exceptionally large group stretches across an appreciable fraction of the solar disk.

At first glance the umbra of a mature spot appears a uniform dark brown or black. Yet if conditions are favorable, careful examination with moderate to high magnifications may reveal the presence of minute bright spots. These features, called umbral dots, vary considerably in brightness and are not always visible.

More complex than a sunspot's umbra is its delicate, weblike penumbra. This structure consists of nar-

row, bright filaments on a darker background extending outward from the umbra. The umbra-penumbra border is usually well-defined but often irregular; look for points of umbral material invading the inner regions of the penumbra. Occasionally some of these projections will reach all the way across the penumbra to the photosphere.

Perhaps the greatest spectacle a solar observer can hope to witness is the eruption of a flare. In a matter of seconds an active solar region can release an astonishing amount of energy. Ordinarily such eruptions originate either in the solar chromosphere or the corona, and an H-alpha or other narrow-band filter is needed to see them. On rare occasions, however, the activity extends down to the level of the photosphere. When this happens the sudden release of energy can radiate across the entire electromagnetic spectrum, producing the rarest of all photospheric events, a white-light flare.

White-light flares are associated with complex, well-developed sunspot groups, particularly those containing several spots within a single penumbra. They appear as sudden, rapidly expanding areas of brightness, and are most often seen when projected against the dark background of a sunspot. A flare usually reaches peak brilliance in just a few minutes, after which it diminishes gradually in intensity.

The break up of a large sunspot group is a gradual process. First to disappear are the small spots scattered between the leader and the follower. Next the follower separates into several smaller spots, and these too gradually fade away. The leader, which has been growing progressively rounder and smaller, is the last to vanish. If the group is especially long-lived, the entire process, from birth to final dissolution, may take as long as two months.

With so much to see, it's no wonder solar observers are eagerly watching the high solar activity following the 1990 maximum. If you count yourself a serious stargazer but have never devoted much attention to our nearest stellar neighbor, now is the time to add it to your observing schedule. Remember, the Sun is not only the most dynamic object you can observe — it is also the only star in the universe you can study in detail with a backyard telescope. □

Edmund Fortier is an active observer living in North Scituate, Massachusetts. His last article was "Inside the Hyades" (February 1990).

WHAT MAKES VENUS GO?

"Earth's twin" isn't, by a long shot. It's more a case
of arrested development. One reason may be that
Venus lost its ocean long ago.

By Robert Burnham

All images courtesy JPL/NASA unless otherwise noted

HIGH AND PRECIPITOUS, the cliff at the edge of Dali Chasma (foreground) shines bright where the ground has weathered. The dark area just below the lip is a relatively fresh landslide. (Vertical exaggeration is 10x; missing map tracks have been filled by interpolation.)

L ast September the Magellan spacecraft completed its third radar mapping cycle of Venus. About 99 percent of the planet is now imaged. At the same time planetary scientists met at the California Institute of Technology in search of answers to major questions regarding the planet: What is the "geological style" of Venus? How does it work? Why doesn't Venus have Earth-like plate tectonics? What has been the planet's history?

Heat appears to be the key to understanding Venus. All planets retain heat left from their formation and from the decay of radioactive elements. On Earth, heat flow drives a process called plate tectonics, which creates new crust. Scientists expected Venus to follow suit because it is virtually Earth's twin in size.

Earth's crust is divided into many plates. Each is a rigid piece of basaltic rock bounded by long narrow zones (such as the Mid-Atlantic Ridge) where molten basalt oozing up from the mantle pushes the plate sideways. After millions of years the displaced crust reaches an ocean trench such as the western coast of the Americas or Japan where it buckles downward and slips beneath a continental edge. Returning to the mantle, the crust eventually melts and is re-erupted.

Venus, however, has no such tidy cycle. "To understand Venus, start by imagining Earth with plate tectonics stalled out," says Sean Solomon of the Carnegie Institution of Washington. Thanks to Magellan, Venus is now better mapped than Earth (where fully detailed charts of the ocean basins remain military secrets). It's clear, though, that Venus has no plates like those of Earth.

In fact, some scientists think Venus became a one-plate planet when its surface got *too* hot. On Earth, the cold surface temperature helps keep oceanic crust rigid. But if Venus started with an ocean, it is long gone and the surface is now halfway to the melting temperature of rock. Its crust therefore is ductile. Like a giant "soft batch" cookie that's ever pliable, the Venus crust isn't stiff enough to slide around in large, intact pieces. Instead, when subjected to forces from within, the Venusian crust wrinkles and puckers largely in place.

"It's likely that Venus is actively building mountains at present," says Solomon. So far scientists have seen no clear evidence for surface changes during the mission, he says, but "high mountain regions like Maxwell Montes will collapse under self-gravity in about 10 million years." Either

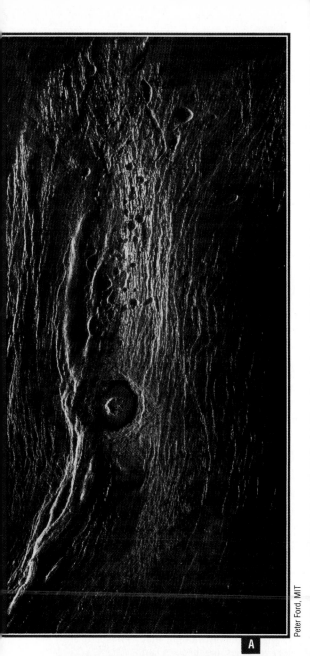

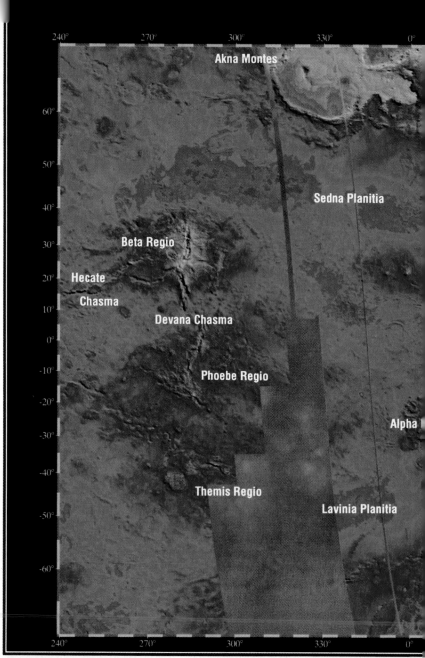

Akna Montes

Sedna Planitia

Beta Regio

Hecate
Chasma

Devana Chasma

Phoebe Regio

Alpha

Themis Regio

Lavinia Planitia

A

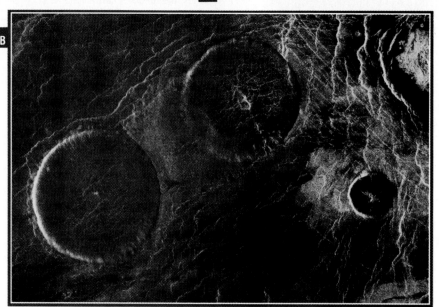

B

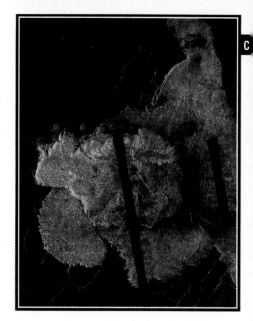

C

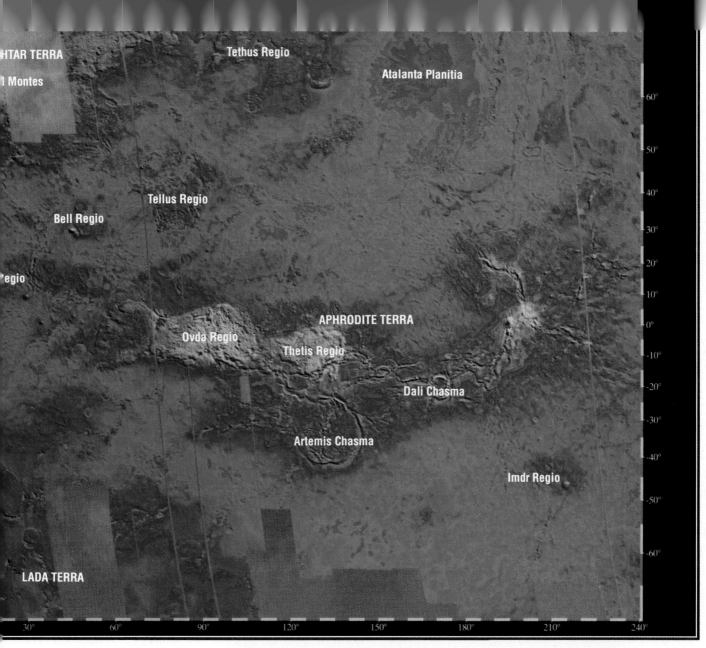

HTAR TERRA

Montes

Tethus Regio

Atalanta Planitia

Tellus Regio

Bell Regio

egio

APHRODITE TERRA

Ovda Regio

Thetis Regio

Dali Chasma

Artemis Chasma

Imdr Regio

LADA TERRA

60°
50°
40°
30°
20°
10°
0°
-10°
-20°
-30°
-40°
-50°
-60°

30°   60°   90°   120°   150°   180°   210°   240°

THE NEW FACE OF A WORLD appears in Magellan's radar map (above), which has many times the detail of the similar map created by Pioneer Venus a decade ago.

A: A DIRECT HIT on the Akna mountain belt. Lava flooded this 22-kilometer crater, but the lack of subsequent deformation shows that the forces that folded the mountains have been quiet since the impact.

B: STIFF AND THICK AS BATTER, the lava that made these 65-km-diameter pancake domes was more viscous than most on Venus.

C: MULTIPLE ERUPTIONS show in this volcanic structure with a beautiful "fan" of lava that's over 100 meters thick at one edge.

D: RISING MAJESTICALLY and draped in fresh (dark) lava flows, Maat Mons stands 8 km high. (Vertical exaggeration 10x.)

D

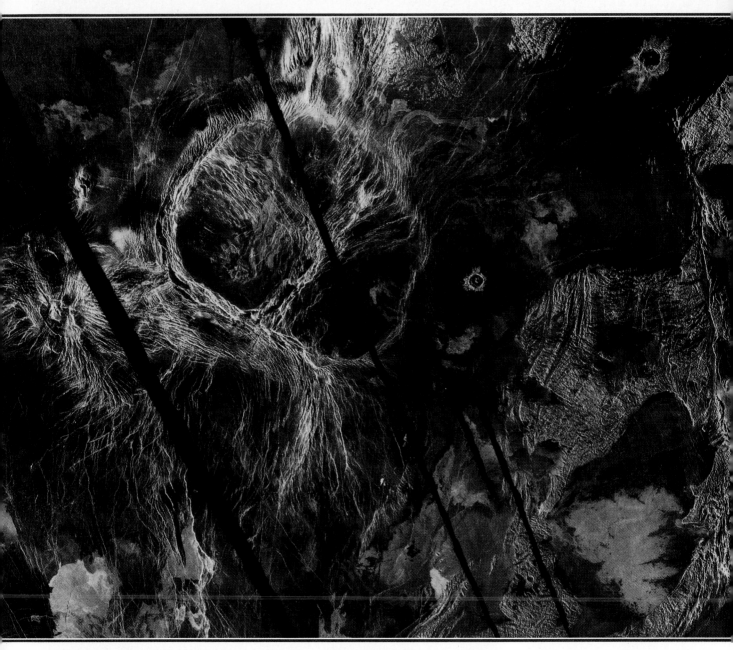

A GEOLOGICAL TANGLE, Lada Terra (above) contains many kinds of features, from small impact craters to the giant oval features called coronas, which are found only on Venus. Eithinoha corona (upper left) is 500 km across.

A: TWO ODDLY SHAPED DOMES lie in Beta Regio; the larger is 45 km across and has steep interior slopes.
B: CRATERS FILLED WITH LAVA are rare on Venus because the surface is largely inactive; Heloise crater (bottom center) is 40 km across. Old, greatly deformed terrain shows bright in the upper part of the image.
C: A BRIGHT APRON OF DEBRIS surrounds Dickinson crater, 69 km in diameter, which appears to have been flooded in two or more stages.

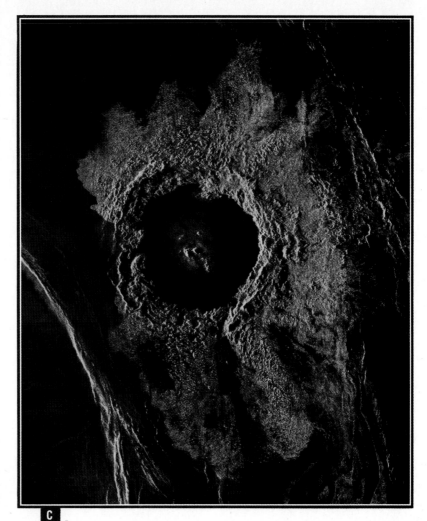

C

Magellan caught Venus at a lucky moment or mountain-building is continuing.

The small number of craters tells Venus scientists that nearly all the present surface is no older than about 500 million years old. That's about twice the average age of Earth's surface, but eight times younger than the Moon's. Researchers disagree sharply, however, over how it came to be that old.

Gerald Schaber of the U. S. Geological Survey is one of many who think resurfacing occurred in something of a spasm that trailed off abruptly about 500 million years ago. Roger Phillips of Washington University in St. Louis represents a smaller number of scientists who believe resurfacing continues intermittently today. Both sides agree it will take a lot of careful mapping to settle the matter.

Donald Turcotte of Cornell University has proposed that resurfacing could run in cycles. Periods like today, with no plate tectonics, would alternate with active periods. As the plate tectonics conveyer belt moved hot mantle material to the surface, it would rob internal heat that makes the process go. With a cool interior, plate tectonics would shut down.

Then heat from radioactive decay would start to build again, until it was high enough to trigger another onset. Turcotte thinks this off-and-on cycle would continue until the interior cooled too far to restart it.

Magellan is now in its fourth orbital cycle. Scientists are carefully tracking it to map Venus' gravity field for clues to the planet's interior structure. However, the craft confronts an uncertain future, both at Venus and on Earth. Its transmitters are overheating and can no longer make radar images. Back here, Magellan faces a different kind of heat: the federal budget. With Mars Observer and Galileo en route, the Cassini Saturn orbiter in development, and many exciting new projects planned, NASA is skeptical about spending already tight funds to extend Magellan's mission past May 15, the end of cycle 4.

If mission planners do get an extension, they will dip Magellan into the upper atmosphere. This would produce a lowered circular orbit and greatly enhanced gravity measurements. However, the price tag for extending the mission is $30 million, and the project's scientists are glumly aware that this money may be more needed elsewhere. □

THE COSMIC ORIGINS OF
LIFE ON EARTH

by Christopher Chyba

The essentials of life may have arrived from space billions of years ago and set the stage for life on early Earth. Because of the hostile environment, life may have had to start many times before winning a foothold.

Adolf Schaller

SOLAR SYSTEM BODIES like Comet Halley (left) and the Murchison meteorite (below) contain organic molecules, the stuff from which life is constructed. The Murchison meteorite, found in Australia, is rich in simple amino acids, the building blocks of proteins.

Bobby Bus/Palomar Observatory

J. William Schopf/UCLA

P icture a Full Moon in the night sky. Its bright areas are mountainous highlands; its dark areas are comparative- ly smooth lowland plains called maria, Latin for seas.

If you gaze at the maria for just a little while, you can quickly convince yourself that many of them are circular in shape. Surprisingly, the reason they are circular is wrapped up with one of our deepest mysteries: the origin of life on Earth. The standard textbook story of how life arose is that simple organic molecules in the oceans of early Earth combined, slowly, to form molecules of greater and greater complexity. These eventually formed a molecule that could copy itself, and the evolution of life was off and running.

However, recently scientists have taken a serious look at an alternative hypothesis for the formation of life — that or- ganic molecules were carried to Earth from extraterrestrial sources such as asteroids, meteorites, comets, and inter- planetary dust particles, and that the origin of life got a boost from these cosmic sources. Moreover, it seems likely that life may have started more than once — perhaps nu- merous times — and because of the hostile conditions on early Earth, repeatedly died out, only to start up again. To examine the credibility of these ideas, let's look at the con- ditions that existed on early Earth and in the adolescence of the solar system as a whole. This requires a bit of a detour from organic molecules bonding into self-replicating sys- tems. Instead, we must learn about the impact cratering record of Earth and the Moon. It's an unusual place to start in what becomes a very unusual journey of discovery.

Raining Rocks on the Moon

Many of the lunar maria are circular because they got their start as impact basins — essentially giant craters left when asteroid-sized objects struck. (The largest basin on the nearside, Imbrium, is nearly 1,200 kilometers in di- ameter.) Later, after the basins formed, dark lava flows seeped in and filled them. The lava naturally sought the lowest terrain, but the impacts themselves may have played a key role by excavating far enough into the Moon to give lunar magma a route to the surface. The resulting maria are dark for the same reason that lava flows on Earth are dark once they cool — volcanic basalts reflect little light.

The rocks returned from the Moon by the American Apollo and Soviet Luna missions allowed scientists to build a picture of the lunar surface, and to determine when the critical events occurred. Consider, for example, the Apollo 12 landing site in Oceanus Procellarum, the Sea of Storms. The astronauts made a number of traverses of the area, collecting a variety of samples. These samples came back to Earth, where scientists determined their ages using radioactive dating techniques.

All told, the Apollo missions returned a total of 380 kilograms — 840 pounds — of rocks. Most of the rocks from a particular mission have ages that cluster around the same value — for Apollo 12, about 3.2 billion years old. This means that the particular ancient lava flow on which the Apollo 12 astronauts landed solidified at this time in the past.

Knowing the age of this particular flow, scientists can go to pictures taken from orbit and count how many craters it has. Geologists can be certain that all these craters were produced after that flow solidified, because any craters formed prior to that time were "wiped clean" from the lunar surface, buried by the flowing lava.

Obviously planetary scientists can't count craters of all sizes. For craters that are too small, researchers can't

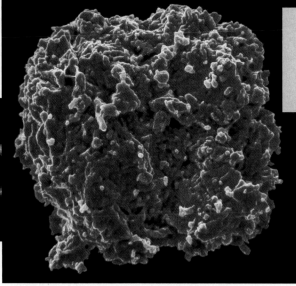

INTERPLANETARY DUST PARTICLES contain about 10 percent organic molecules and fall slowly to Earth, contributing 300 tons of organics per year. Although each spans less than a tenth of a millimeter, in Earth's early history they may have been a key source of organic molecules.

John Bradley/MVA Associates and Don Brownlee/University of Washington

discern them clearly, even from as close as lunar orbit. In practice, lunar crater counters usually try to pick out only craters bigger than 4 kilometers in diameter. The data from all of the Apollo and Luna missions (see the diagram on page 33) show that prior to about 3.5 billion years ago, the level of cratering was much higher than today. But since then, cratering has been roughly constant at a comparatively low level. The intense cratering prior to 3.5 billion years ago is often referred to as "the heavy bombardment."

The heavy bombardment is consistent with astronomers' current models of solar system formation, which say that as the planets grew in size, they either slowly swept up or gravitationally scattered asteroid and comet-sized bodies called planetesimals. Some planetesimals spun into rogue orbits deep in the inner solar system, and some eventually collided with one of the inner planets. As the planets grew bigger and bigger, more and more planetesimals were swept up or scattered away, and the supply of planetesimals declined. Finally, by 3.5 billion years ago only a very small number of comets and asteroids were left orbiting through the inner solar system and the impact rate dropped sharply as the population thinned.

Lessons from Mercury and Mars

Can we be confident that this is the right story? How sure are we that the objects that excavated the basins and craters on the Moon didn't hit just the Moon and miss Earth altogether? In fact, the Moon probably started its life only a few terrestrial radii away from Earth, evolving outward to its current distance of 60 Earth radii. We might worry, for example, that as the Moon moved out and away from the Earth, it swept up a swarm of debris that dug out the lunar craters. If this were the case, cratering on the Moon would tell us nothing about cratering on Earth. However, spacecraft missions to Mercury and Mars have helped resolve this issue. Start with the oldest parts of the Moon, and imagine counting up the number of craters of different diameters. On the Moon, you find that when you go down a factor of ten in crater size, the craters become more common by about a factor of a hundred. Of course this rule isn't perfect, and some crater sizes are present in greater or lesser number than this simple rule leads you to expect.

Now play the same game with craters on the ancient terrain of Mars, or on Mercury, and what do you find? Not only do you find the same overall relationship between crater number and crater size, but those particular sizes

that broke the rule on the Moon break the rule to about the same extent on Mars and Mercury as well. A common interpretation of this similarity in cratering records is that all these worlds were cratered by the same population of objects, a population that had a certain overall size distribution with certain particular sizes over- or under-represented. But if Mars, Mercury, and the Moon were all pummeled by the same population of impacting objects during the heavy bombardment, Earth and Venus must have been as well.

The problem with Earth (and Venus) is that it is so geologically active that it no longer has any of its most ancient crust left intact. We know from radioisotope dating on certain meteorites that Earth is about 4.5 billion years old. Nevertheless, almost no surfaces remain on Earth that date from before 3.5 billion years ago. Not enough ancient surface is left to have preserved a record of the heavy bombardment. On Earth, and Venus, that information has been erased. But it has been retained on at least parts of the surfaces of the Moon, Mercury, and Mars.

Thus, if we want information about the environment of the early Earth, there is no way to get it from terrestrial geology. We simply have to look to the Moon and planets, and try to extrapolate from them back to our own world. To learn about Earth, we must look beyond it. This is true not just in some metaphorical sense, but is quite literally true.

Enter Life

We face the same dilemma when trying to unravel the development of early life. The oldest rocks on Earth capable of having fossils in them have fossils. These include microscopic fossils of bacteria, and the fossilized remains of macroscopic structures, called stromatolites, formed by mats of algae. The sedimentary rocks containing these remains are found in Western Australia and South Africa and are 3.5 billion years old. There are older rocks (3.8 billion years) in Greenland, called the Isua formation. The Isua rocks began as sediments, but over their long life they have several times experienced high temperatures and pressures, making it unlikely that any fossils could have survived within them. Still, controversial claims have been made for signs of life in the Isua rocks as well. The problem is that the Isua formation has been so reworked that it is difficult for scientists to know what they are seeing. Finally, newly discovered granitic rocks found in Canada's Northwest Territories are about 4.0 billion years old. But these rocks formed within Earth at high temperatures and pressures, and so are incapable of containing any fossils.

Thus no record exists of the origin of life on Earth. The oldest geological record already contains fossils of sophisticated, probably photosynthesizing microorganisms. So scientists can't get much information about the terrestrial environment before or during the time of the origin of life by looking at Earth. Yet we know from the Moon that this environment must have been an extremely violent, impact-ridden one. The earliest terrestrial fossils coincide in time with the last stages of the heavy bombardment that we have dated from the lunar samples. It seems unlikely that this similarity of dates is just a coincidence. Rather, the heavy bombardment probably played important roles — both positive and negative — in the timing and origin of life on Earth.

But as with any area of research where the data are limited, some scientists contest this conclusion. A minority of lunar scientists argue that the lunar bombardment history presented here is badly mistaken. These scientists note that lunar rocks seem to contain no melted remnants from impacts older than about 3.9 billion years — yet samples of much older lunar lavas and ancient crust have been found. They suggest this means that the Moon experienced only a light bombardment during its first 600 million years, followed by an intense cataclysmic bombardment that produced virtually all of the familiar lunar features.

These conclusions have in turn been disputed. First, the cause of such a "cataclysm" remains a mystery. Second, the lack of ancient impact melts may simply be because, prior to 3.9 billion years ago, the bombardment was so intense that ancient melts were being remelted, and their earlier formation dates were obliterated. But even if the "cataclysm" hypothesis is correct, the question remains: were there important connections between early impacts and the origins of life?

Janus

Janus was the Roman god known as the doorkeeper of heaven. Since every door looks two ways, Janus had two faces. As a solar deity, he had one face for the rising, and one for the setting, Sun. It was natural, then, for Janus to be the patron of beginnings, and the patron of ends.

The heavy bombardment on early Earth had this Janus-like quality. It may have repeatedly exterminated life, but it may also have been essential for the origin of life. However, let's first turn to the destructive side of the heavy bombardment, the face of Janus for the setting Sun.

Basin-making impacts on early Earth would have blown some of the atmosphere off into space. Because a thick atmosphere is essential for keeping a planet warm, such "impact erosion" of an atmosphere could have converted an initially hospitable planet into a freeze-dried desert. In fact, many planetary scientists now think this might be exactly what happened to Mars, which once had flowing water on its surface, but now seems to be a nearly airless, dead world.

Apparently Earth's larger size saved it from the same fate. Earth's larger gravity (over $2^{1}/_{2}$ times stronger) makes it correspondingly harder to blast portions of its atmosphere off into space. Detailed studies show that on Earth, asteroids and comets delivered more atmospheric gases than they blew away into space. On Mars, the opposite was probably the case. In the case of the Moon, gravity is so low that impacts would slowly erode not the Moon's atmosphere (which is virtually nonexistent), but the Moon itself. The Moon is slowly being blasted away into the cold confines of space.

Especially big impacts on early Earth had a far more serious effect than just blowing gases into space — including gases helpful to the origin of life. Any one of the largest impacts would have produced a shortlived global atmosphere composed of rock vapor, temporarily raising the temperature of Earth's surface to above that of the inside of an oven. In the most extreme cases, this searing heat would have lasted long enough to have evaporated the entire terrestrial ocean, sterilizing the surface of Earth.

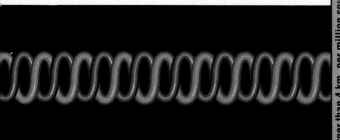

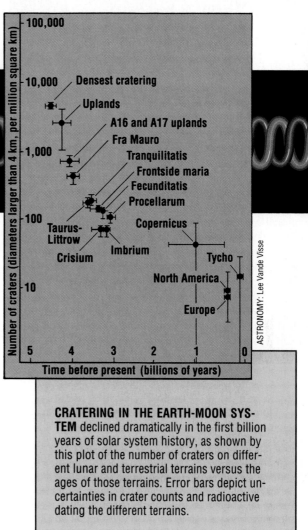

CRATERING IN THE EARTH-MOON SYSTEM declined dramatically in the first billion years of solar system history, as shown by this plot of the number of craters on different lunar and terrestrial terrains versus the ages of those terrains. Error bars depict uncertainties in crater counts and radioactive dating the different terrains.

Scientists can use the cratering record on the Moon to estimate just how often this level of destruction took place. Statistically, because of Earth's larger gravity, something like 17 or so objects larger than the largest object that hit the Moon should have collided with Earth. If the largest object that impacted the Moon was the one responsible for the 2,500-km-diameter South Pole-Aitken basin on the lunar farside (whose controversial existence was finally confirmed two years ago by the Galileo spacecraft), Earth was probably hit about five times by asteroids or comets big enough to have completely vaporized its oceans. Another way of saying this is that Earth's surface should have been completely sterilized some five times or so.

If this is right, then life on Earth did not have a billion or so years to get started, the time between Earth's formation and the first recorded fossil life. Life may have had only a tenth of this time. And if life got started this quickly, it probably means that the origin of life was a very easy process.

The dilemma, of course, is with the estimate "five times or so" for the number of times that giant impacts might have heat-sterilized Earth. It could be that Earth's surface was sterilized not 5, but 10 times, and it's not out of the question that it never happened at all. Scientists can't resolve this issue with certainty because there's no record left on Earth. All they can do is statistically extrapolate from the Moon. This puts them in the frustrating position of having to admit that the most calamitous events in Earth's history (after the giant primordial impact that many planetary scientists now suspect was responsible for forming the Moon) can be spoken of only in terms of probabilities, in terms of might-have-been's, or probably-were's.

In any case, even if Earth's surface was never entirely sterilized, many smaller but still giant impacts would have wiped out this or that "warm pond" where promising evolutionary experiments in the origin of life were underway. Somehow life emerged successfully in the midst of all this impact violence.

One way around this dilemma is for life to have evolved not at Earth's surface but at comparatively protected locations deep in the oceans or underground. If life could have evolved, say, at deep-sea hydrothermal vents, its earliest evolution would have been protected against all but the most devastating impacts. But vehement disagreements have raged among origins-of-life scientists over whether hydrothermal vents are in fact suitable locations for the sorts of reactions thought necessary to get life started.

Another possibility is that life evolved at Earth's surface but quickly expanded to deep ocean or underground environments. Surface life could then have been obliterated by impacts, leaving only the subsurface communities of microorganisms intact. Life on Earth today would then be the descendants of an early subsurface ecology, simply because these were the only organisms that made it through the heavy bombardment.

The Rising Sun

But Janus had two faces, contemplating not only the sunset but also the sunrise. And at the same time that comets and asteroids were wreaking devastation upon the surface of Earth, they may also have been setting the stage for the genesis of life.

Comets and some asteroids are rich in exactly those elements needed for the origin of life. For example, comets are about 50 percent water ice by mass. If ten percent of the objects that collided with Earth were comets, all our planet's oceans can be explained. Because liquid water is essential to life as we know it, it could well be that comets and water-rich asteroids thereby "fertilized" Earth's surface, rendering it hospitable for the origin of life. And just as important, it appears that much of the terrestrial inventory of biologically critical elements, such as carbon and nitrogen, may have also been delivered in this way.

J. William Schopf/UCLA

NASA

APOLLO 12 ASTRONAUTS failed to detect life on the barren surface of the Moon. Yet the lunar cratering record allows astronomers to reconstruct the period of heavy bombardment and what it must have done to Earth. The oldest known microfossil (above) is 3.4 to 3.5 billion years old, dating back to about the time of the heavy bombardment.

Moreover, comets and some asteroids are rich in organic molecules — those molecules out of which all life-as-we-know-it is made. It's possible that the organic building blocks of life on Earth were initially brought to Earth from outside. In fact, these processes continue to some extent today. Some meteorites, such as the Murchison meteorite (so called because it fell near Murchison, Australia), contain various interesting organics. For example, Murchison is rich in amino acids, the building blocks of proteins. Far more extraterrestrial organics, however, are falling to Earth in the form of interplanetary dust particles (IDPs). These particles, smaller than a tenth of a millimeter across, are small enough to decelerate completely high in the atmosphere before they are heated sufficiently to burn up. They settle slowly and gently down to Earth's surface. As a consequence, their organics — IDPs appear to be about 10 percent organic by mass — survive delivery to the surface of Earth.

Earth is currently collecting some 300 tons a year of extraterrestrial organics delivered in IDPs. Although individual meteorites are vastly larger than any IDP, falls of organic-rich meteorites are so rare that meteorites are now providing Earth with less than 10 kilograms of organic matter a year — less than one ten-thousandth as much as IDPs provide. Nevertheless, comparatively big falls like Murchison are essential, because they provide cosmochemists with chunks big enough to work with.

There are, of course, more dramatic ways to bring organics down to Earth. Over 130 craters are now recognized on Earth's surface, nearly all representing impacts by comets and asteroids over the past few hundred million years. (Most older craters have long since eroded away or been filled in.) Although big impacts with Earth are, fortunately for us, quite rare, far more mass arrives on Earth in these intermittent events than in the slow, steady accumulation of IDPs. If some of these big impactors were organic-rich comets or asteroids, Earth could potentially accumulate vast quantities of organics from this source.

But despite its importance at first glance, this mecha-

nism seems not to work. It is easy to see why: collisions of big asteroids or comets with Earth are extremely violent events. If a body the size of Halley's Comet hits Earth, the resulting explosion has the energy of 100 million tons of TNT. It is difficult to see how organics could survive this kind of holocaust. Detailed computer simulations of the destruction of organics in small comets show that only small comets (hundreds of meters across, say) might have successfully delivered organics to early Earth, because only these bodies are small enough to be slowed sufficiently by the atmosphere for their organics to survive the collision. It's likely that the amount of organics collected by Earth in this way was outweighed by those brought in by IDPs.

Extraterrestrial vs. Terrestrial Sources of Organics

How important were the organics brought in from out there for the origin of life down here? Because scientists don't really understand the origin of life, this question can't be answered with confidence. One way to approach the issue, though, is to compare the extraterrestrial sources with likely sources of organics made on Earth. Since the early 1950s, scientists have worked out a number of ways for making organics via chemical reactions in Earth's atmosphere. How do these stack up against the fine rain of organic-rich dust from the sky?

The answer to this question is complex, because the production of organics on early Earth depends on the details of Earth's primitive atmosphere. Once again, scientists are hindered by a lack of data. About forty years ago, when the first key experiments on organic production in early terrestrial atmospheres were performed, scientists thought Earth's earliest atmosphere consisted of methane and ammonia. In this kind of atmosphere, organics are produced copiously from lightning, ultraviolet light, and other sources. Organic production from these sources is then so efficient that any direct importation of extraterrestrial organics is completely swamped by terrestrial sources.

However, even in this case the heavy bombardment

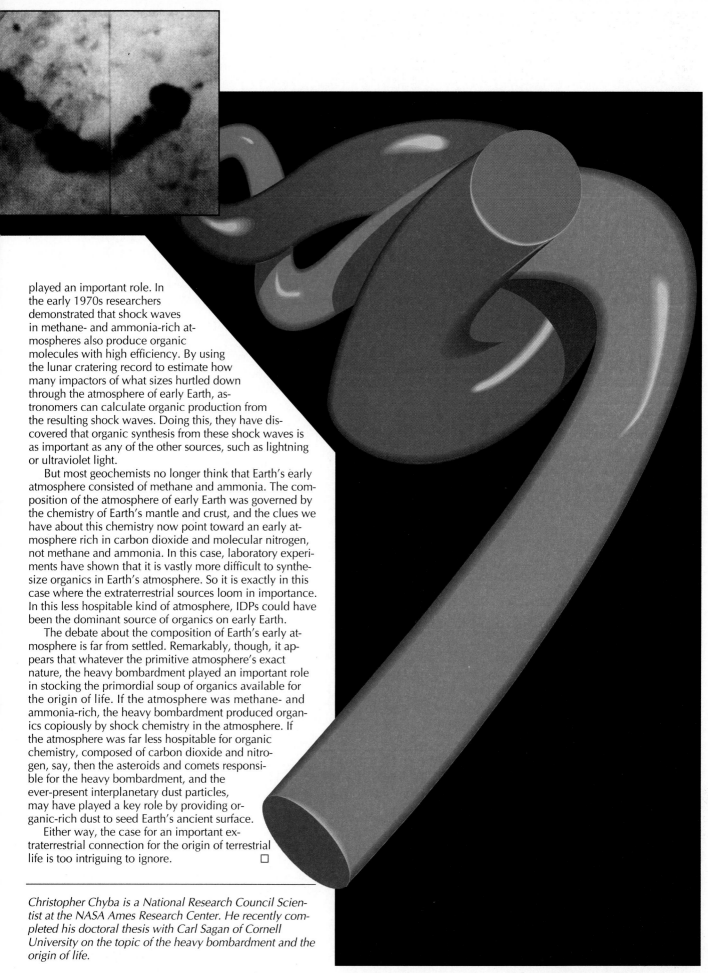

played an important role. In the early 1970s researchers demonstrated that shock waves in methane- and ammonia-rich atmospheres also produce organic molecules with high efficiency. By using the lunar cratering record to estimate how many impactors of what sizes hurtled down through the atmosphere of early Earth, astronomers can calculate organic production from the resulting shock waves. Doing this, they have discovered that organic synthesis from these shock waves is as important as any of the other sources, such as lightning or ultraviolet light.

But most geochemists no longer think that Earth's early atmosphere consisted of methane and ammonia. The composition of the atmosphere of early Earth was governed by the chemistry of Earth's mantle and crust, and the clues we have about this chemistry now point toward an early atmosphere rich in carbon dioxide and molecular nitrogen, not methane and ammonia. In this case, laboratory experiments have shown that it is vastly more difficult to synthesize organics in Earth's atmosphere. So it is exactly in this case where the extraterrestrial sources loom in importance. In this less hospitable kind of atmosphere, IDPs could have been the dominant source of organics on early Earth.

The debate about the composition of Earth's early atmosphere is far from settled. Remarkably, though, it appears that whatever the primitive atmosphere's exact nature, the heavy bombardment played an important role in stocking the primordial soup of organics available for the origin of life. If the atmosphere was methane- and ammonia-rich, the heavy bombardment produced organics copiously by shock chemistry in the atmosphere. If the atmosphere was far less hospitable for organic chemistry, composed of carbon dioxide and nitrogen, say, then the asteroids and comets responsible for the heavy bombardment, and the ever-present interplanetary dust particles, may have played a key role by providing organic-rich dust to seed Earth's ancient surface.

Either way, the case for an important extraterrestrial connection for the origin of terrestrial life is too intriguing to ignore. □

Christopher Chyba is a National Research Council Scientist at the NASA Ames Research Center. He recently completed his doctoral thesis with Carl Sagan of Cornell University on the topic of the heavy bombardment and the origin of life.

LUNAR GEOLOGY

Mysteries of
the Moon

There's an unknown Moon hanging silently, beautifully in our
sky. It's the Moon whose mysteries Apollo didn't fully fathom
— and it may be the most intriguing Moon of all.

by Damond Benningfield

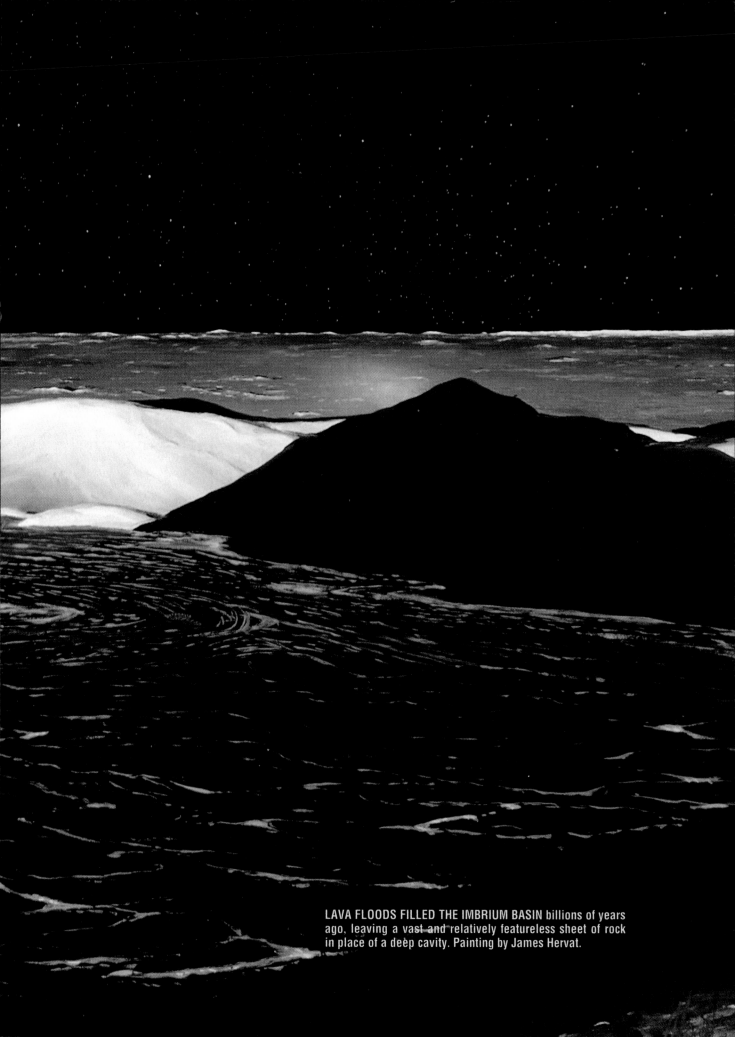

LAVA FLOODS FILLED THE IMBRIUM BASIN billions of years ago, leaving a vast and relatively featureless sheet of rock in place of a deep cavity. Painting by James Hervat.

Project Apollo was supposed to be the next-to-last chapter of an exciting celestial mystery novel. Earth-bound scientists were supposed to take the 842 pounds of lunar rocks and soil, thousands of sharp photographs, and millions of bits of electronic data and out of them piece together the Moon's origin, its evolution, its structure, and its composition. Just open the boxes, fiddle with the rocks a bit, write up the last chapter, and — *voilà!* — the Moon would be a closed book.

Reality, of course, is seldom so tidy. Data from Apollo and a flotilla of unmanned Moon orbiters, crashers, and landers have yielded many answers, but few of the solutions are as elementary as those in a Sherlock Holmes story.

For example, consider the question of the Moon's origin. Working from Apollo data, planetary scientists now believe that the Moon was born in a violent collision between Earth and a Mars-sized planetoid soon after Earth formed. But this giant impact hypothesis is still little more than a general idea; details are hazy. There is lively debate, for instance, over the exact parentage of the present-day Moon: Did it form from material gouged from the primitive Earth? Or from debris from the rogue planet? Or some combination of the two? For that matter, is our present Moon an only child or the sole survivor of a multiple-moon system?

Or look at the Moon's brutally scarred surface. Billions of years of bombardment by meteorites of all sizes have left a tangled history we understand only partly. Many scientists think the period of violent activity lasted several hundred million years during the Moon's earliest years. But some say the barrage might have taken place during a tight 40-million-year window — perhaps after other moons were pulverized in their own giant collision.

And then there's the issue of lunar water. There isn't any. That is, unless hidden pockets of ice have survived billions of years of harsh conditions at the Moon's poles.

"We have a much more complicated picture of the Moon today than we had in 1974, when a lot of people got out of the field," explains Graham Ryder, a U.K.-born staff scientist at the Lunar and Planetary Institute in Houston. "It's like the Bob Dylan line 'We were so much older then.' We used to know it all."

Today, planetary scientists are quick to acknowledge that they don't know it all. In fact, lunar research is an active and healthy field of study. Researchers continually find new ways to dissect the Apollo samples and to interpret old results. And new

APOLLO POINTED THE WAY toward the solution of many lunar questions, but it couldn't provide the exact answers planetary scientists now seek.

generations of ever-faster supercomputers make it possible to simulate a wide range of conditions throughout the Moon's lifetime. But even with our new tools, new talent, and a libraryful of new ideas, the Moon continues to hold tight to some of its oldest and most cherished secrets.

How to Make a Moon

One of the most intriguing secrets — its solution was a sort of Holy Grail for Apollo — is the Moon's origin. For many years scientists thought the Moon was born in one of three ways: it formed from the same primordial cloud of gas and dust as Earth, or it spun free from the primordial Earth, or it was a stray that wandered too near Earth and was captured by the larger body's more powerful gravitational pull. But the Apollo samples demonstrated that the correct answer is "none of the above."

"The three classical theories all falter on the chemistry," says Jay Melosh, a planetary scientist at the University of Arizona's Lunar and Planetary Laboratory in Tucson. Melosh notes that the giant-impact hypothesis, first suggested in the mid-1970s but not gaining wide acceptance until a decade later, "does a pretty good first-order job of explaining the chemistry and dynamics of the Moon." For example, it explains why the Moon lacks heavy elements common on Earth.

According to Melosh's model, very soon after Earth formed 4.5 billion years ago, a planet about 1.5 times more massive than Mars (or 15% as massive as Earth) struck the newborn Earth at an oblique angle. The impact vaporized a large part of Earth's crust and upper mantle, spewing it into space like a jet. The blast also accelerated Earth's rotation, spinning the planet up like a child's top.

The jet contained mostly light material from Earth's outermost layers. Largely missing were heavy iron and nickel — the main components of Earth's core. The material that squirted into space settled into a hot disk around Earth relatively nearby. The disk quickly cooled and the vapor condensed to form a thick ring of small particles. In perhaps only a thousand years or less, these particles coalesced to form the Moon.

So Far, So Good

Most planetary scientists agree with this broad outline of events, but there are exceptions. Some believe that the vapor disk was created by a series of smaller impacts. But giant-impact proponents counter that such a sequence could not account for Earth's

SHOWPIECE CRATER OF THE LUNAR FARSIDE, Tsiolkowsky invites scrutiny from those who want to probe its structure, study how its rocks relate to the surroundings, and analyze its lava flows. For example, are all the craterlets on its dark floor caused by impacts — or are some of them volcanic?

rapid rotation rate. Multiple impacts would follow a random pattern, and their rotational effects would tend to cancel one another out.

Yet major questions remain unanswered. At the top of the list is the fate of the planet that caused all of this mayhem. One group, led by Melosh, says that as much as half of the impactor was vaporized in the collision and became part of the disk that later formed the Moon. Another group, led by Alastair G. W. Cameron, an astronomer at Harvard-Smithsonian Center for Astrophysics, says that the Melosh scenario "is possible, but it's probably not the way things really happened." Instead, Cameron believes, most of the impactor skipped off the rattled Earth, like a stone skimming over a pond, with its core sinking through the molten soup all the way to Earth's core.

Calculating Chaos in a Computer

This will be a difficult question to resolve. No one knows the chemical composition of the impactor, but if enough Moon samples turn up elements that can be identified as not occurring on Earth, there's a good chance they came from the object. Yet it may prove impossible to know for sure how much of the impactor was incorporated into the Moon and how much merged with Earth. Both Melosh and Cameron are using supercomputers to run complex simulations of conditions during and after the impact. These help narrow the range of possibilities, though they are unlikely to provide definitive answers.

So far, the simulations have focused on the impact itself and on the first chaotic minutes after. No one has even tried to model the details of lunar accretion — the process by which tiny lumps of matter merge to form larger lumps and eventually a Moon. Ryder thinks that when someone does model the process, the results will be astounding — the cooling disk will yield several Moons.

"My own gut feeling," he says, "is that Earth had three or four Moons." Ryder bases this on an examination of meteorite bombardment of the lunar surface a few hundred million years after the Moon formed.

Although many researchers argue that the Moon was blasted by meteorites for several hundred million years, Ryder says that Apollo samples indicate there was little meteorite activity from about 4.4 billion years ago — when a primitive solid crust formed atop the molten interior — until 3.9 billion

FAMILIAR TO EVERY TELESCOPE OWNER, Copernicus was to be a target for one of the canceled Apollo missions. Moon scientists think this 810-million-year-old impact could help us probe the layers underlying Mare Imbrium.

years ago, "when all hell broke loose." Then from 3.87 to 3.83 billion years ago, Ryder believes the Moon underwent a "cataclysmic" bombardment, with meteorites pounding the surface mercilessly. The source of this big-league hailstorm? Earth's other Moons, which were pulverized in their own giant impacts, Ryder says.

For evidence, he points to a detailed study of the ages of several Apollo samples. Using a new laser dating technique, Ryder and a colleague have evaluated about a half-dozen samples from Apollo 15, which landed on the rim of the Imbrium Basin, a giant impact feature in the Moon's northern hemisphere. The dating technique compares the ratios of several isotopes (different nuclear forms of a chemical element) inside the sample to determine its age.

The internal clocks of all of the samples were "reset" — something that can be done only when the rocks melt completely — less than 3.9 billion years ago. Ryder says that a leisurely impact rate spread out over a few hundred million years wouldn't generate enough heat to reset the isotopic clocks. "The idea that you can reset these isotopes

without a cataclysm just isn't on," he says. "It's fairly convincing that there was a cataclysm. If we go back to the Moon, we can figure it out."

If Not Green Cheese, Then What?

There are many other questions that scientists don't believe they can answer until astronauts or spacecraft return to the Moon. For example: What is the Moon made of?

Scientists think they have a pretty good general idea of the Moon's chemical composition, but this is based mainly on detailed study of just nine spots on the Moon's surface (six Apollo sites and three unmanned Soviet Luna landing sites). Instruments aboard orbiting Apollo service modules added some rough estimates of the composition of other areas, but none of the instruments scanned more than 20% of the lunar surface, and most saw only about 5%. Our picture of the Moon, unfortunately, is painted with a broad brush — and Apollo taught us that the most interesting secrets are revealed in the micro-details.

"We don't know that much about fine-scale chemical distribution on the Moon," says John Dietrich, curator of the Lunar Curatorial Facility, the Apollo sample repository at NASA's Johnson Space Center in Houston. "We can take a first guess based on what we have. We suspect that the rocks at the poles are similar to those at the equatorial regions, but we just don't know for sure."

All but one of the Apollo missions and all three Soviet Luna sample-return probes visited the lunar maria, the prominent dark "seas" that are vast basins flooded with dense volcanic rock. Only Apollo 16 sampled the highlands regions, the light-colored areas that are mangled collections of mountains, canyons, and craters.

"Most of the Moon, including almost all of the farside, is highlands and most of our samples are from the maria," says Dietrich.

Just as important, all nine sample sites lie within a relatively small box on the Moon's nearside. "We sampled very few kinds of rocks and it's important to sample others," says William Muehlberger, a University of Texas geology professor and principal investigator for the Apollo 16 and 17 geological explorations.

Another problem is that all the samples came from only the top few feet of the Moon's surface. Some of the rock and soil samples were gouged from greater depths by meteorite impacts, but no one can say for certain which samples came from what depth. Until future explorers penetrate well below the surface or obtain samples that scientists are certain were blasted from a given depth, it will be almost impossible to determine the composition (and structure) of the Moon's subsurface layers.

A RETURN TO THE MOON is necessary if we wish to understand fully the history of our closest celestial neighbor — and that of our own planet.

Waters of the Moon?

Perhaps the most surprising gap in our knowledge of the Moon is the lack of detailed maps of the farside and the polar regions. "We have better coverage of Mars than of the Moon," says Ryder.

Five American Lunar Orbiter spacecraft, sent to scout for good Apollo landing sites, snapped about 2,000 sharp photos of the Moon during the mid-1960s. Astronauts and automated cameras aboard the Apollo missions added thousands more. But the orbits of these vehicles seldom carried them more than 30° north or south of the lunar equator. And because all the Apollo landings were planned for the nearside, the Lunar Orbiters concentrated their observations on the face we see from Earth.

This means nearly all the Moon has been mapped at medium resolution and small areas at high resolution. But missions to other planets have shown that getting sharper pictures means that your understanding changes, even dramatically. For example, one gap in resolution and understanding emerged when the Galileo spacecraft swept past Earth and the Moon in December 1990 en route to Jupiter. Images taken by Galileo multispectral imager confirmed the existence of an ancient farside impact basin to the southwest of Mare Orientale. This basin

has been pounded so heavily since its formation that it is almost impossible to recognize as a basin, but it is about 2,200 kilometers across. The feature was not detected in the available orbital photography. The Lunar Observer project, which would place an advanced scientific spacecraft in lunar orbit by the end of the decade, could fill in most of the blanks in our maps of the Moon.

Lunar Observer also could look for something that most scientists don't believe exists — lunar water. Apollo samples have shown no trace of water, past or present. Yet some researchers believe that pockets of water ice and other frozen volatiles might exist at the lunar poles. They say that comets striking the Moon's polar regions might have found a few lucky spots inside steep-walled craters such as Peary, Hermite, and Amundsen that receive little or no direct sunlight. Temperatures during the long lunar day can top 250° Fahrenheit, so any exposure to the Sun would vaporize icy deposits.

At the 1987 Lunar and Planetary Science Conference, James Arnold of the University of California at San Diego suggested that material from comets might become buried after it strikes the surface. In all, he said, as much as 35,000 square kilometers — about 0.1% of the lunar surface — might provide safe haven for icy deposits. Lunar Observer, which will orbit from pole to pole, should detect water ice or other frozen deposits buried up to a meter beneath the lunar surface using highly sensitive gamma-ray spectrometers.

Toward the Unknown Regions

Lunar science today stands on a firm foundation thanks to Apollo. Some people might want to declare the story over, snap the book shut, and go on with other matters. Yet if we stop here we will be abandoning a small planet full of mysteries right on our own doorstep.

The Moon and its history are an integral part of our own planet's past. Apollo pointed out the next questions to ask of the Moon, but our nearest celestial neighbor is too big, too far away, and too old to serve up its secrets without a lot more digging. □

Damond Benningfield is a science writer based in Austin, Texas. His last article for ASTRONOMY was "Where do Comets Come From?" in the September 1990 issue.

DID MARS ONCE HAVE MARTIANS?

Today the Red Planet is cold, dry, and lifeless. But early in its history, conditions were warm and wet — identical to those under which life started on Earth.

by Christopher P. McKay

A MARTIAN YELLOWSTONE PARK could have teemed with microbes some 3.5 billion to 4 billion years ago, when conditions favored the development of life.

MARTIAN VALLEY NETWORKS in the ancient cratered terrain of the southern hemisphere offer the best evidence that liquid water flowed on Mars some 3.8 billion years ago. Thus the planet must have been warmer and had a much thicker atmosphere than today.

A new phase in the exploration of the Red Planet begins on August 24, when the Mars Observer spacecraft arrives there. Our return gives Mars scientists a new shot at an old question, one that has forever intrigued them: Is there life on Mars? Or, just as intriguing, was there *ever* life there? Results from recent studies point more and more toward an early Mars that was astonishingly like the early Earth when life gained a foothold here.

The last successful mission to Mars, the Viking project in 1976, focused in vain on the search for life. Its two landers touched down at distant but similar sites on the northern plains of Mars. They analyzed the composition of the Martian soil and atmosphere, conducted meteorological observations for three Mars years, and most importantly analyzed the Martian soil for evidence of microbial life.

Each Viking lander carried a biology package consisting of three experiments that incubated the Martian soil and looked for signs of metabolic activity. Each lander also possessed a mass spectrometer for identifying organic molecules in the soil. The soil incubated in the Viking biology experiments showed activity which in some ways mimicked what scientists expected from life but which was unlike activity seen in

any soils tested on Earth before the mission. The mass spectrometer did not, however, detect any organic molecules in the soil, even at a level of just a few parts per billion. Because life on Earth literally teems with organics and sheds the stuff everywhere in its environment, it is hard to imagine that life could exist on Mars today with no trace of organics in the soil.

This evidence against life is compounded by a general problem: the absence of liquid water. Temperatures on Mars are so cold and the atmospheric pressure so low that water cannot exist as a liquid on the Martian surface, and liquid water is an essential feature of life on Earth. Thus the lack of organics and the absence of liquid water all but rule out the possibility that life now exists on the surface of Mars.

Looking for Past Life

Conditions on Mars were not always so hostile, however. The most exciting results from Viking on the question of life came not from the biology experiments on the landers but from the cameras on the orbiters. The images returned from Viking and from the Mariner 9 spacecraft that arrived several years earlier showed features that seemed to be clear evidence of copious liquid water at some time in Mars' past.

A BIT OF MARS ON EARTH exists in Wright Valley in Antarctica. Frigid temperatures and almost no precipitation make the valley and Lake Vanda reminiscent of Martian conditions, yet they play host to thriving colonies of algae and lichen.

Two kinds of features seen in the images of the Martian surface point unambiguously to the action of flowing water. One is the outflow channels, large features that appear to be the result of catastrophic outflows of liquid water from water-bearing rock formations, or aquifers, beneath the Martian surface. Planetary astronomers think the outflows were triggered by the melting of ice that was holding back the liquid water. Some of these outflow channels appear to be quite young geologically speaking — less than a billion years old, which makes them particularly intriguing from the point of view of a search for life: Their young age suggests that subsurface, liquid-water reservoirs may still exist on Mars.

The other water-carved features are the valley networks, sinuous valleys that in many cases form highly branched and meandering drainage patterns. Some clearly seem to suggest rain or snow as the source of the liquid water. But rain and snow don't exist in the present Martian climate. Temperatures and atmospheric pressures are too low for any significant precipitation, even snow. Furthermore, if water were to start flowing in the slow, easy way indicated by the networks, it would freeze solid. This distinguishes the networks from the outflow channels, in which the

water flowed with such vigor that it would not freeze even under current conditions.

So it's clear that the climate on Mars was much different at the time the valley networks were carved. It's hard to pinpoint when these valleys formed, but the abundant valley networks on the ancient cratered terrain in the southern hemisphere suggest that the warmer and wetter epoch of flowing water was between 3.5 billion and 4 billion years ago. This straddles the period of late heavy bombardment that occurred 3.8 billion years ago — the time when all the inner planets were subjected to intense cratering from debris left behind from the formation of the solar system.

This was a critical time because we have direct evidence that life on Earth was already well developed in the microbial sense by 3.5 billion years ago. (See "The Cosmic Origins of Life on Earth" by Christopher Chyba, November 1992.) It is hard to find sedimentary rocks older than this age that have not been destroyed by the active nature of Earth's surface. But one set of partially destroyed rocks from 3.8 billion years ago contains evidence that strongly suggests life at this earlier time. Biologists think this life arose in liquid-water environments, but they do not know how long the life-creating process required. From the ages of the

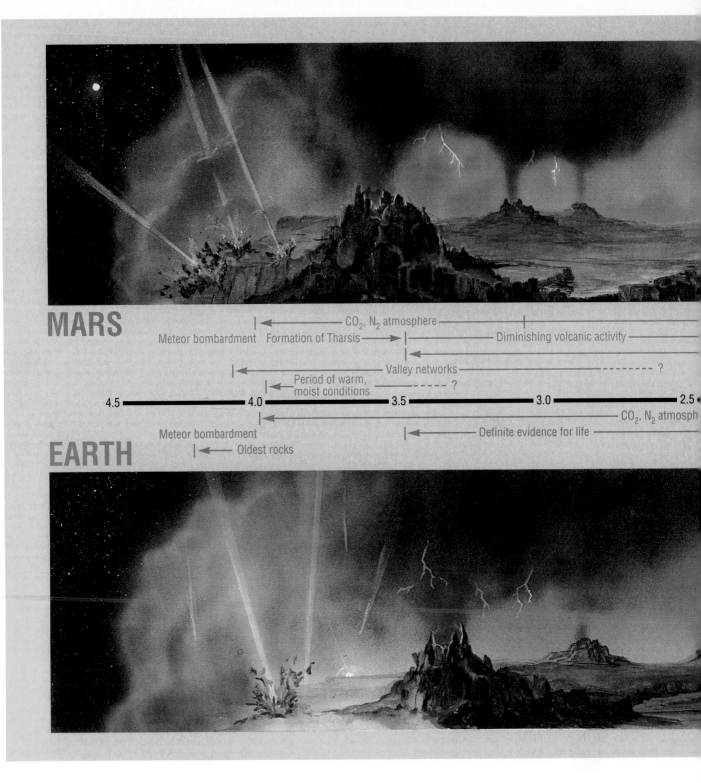

MARS

Meteor bombardment | Formation of Tharsis → | ← CO₂, N₂ atmosphere → | | Diminishing volcanic activity

| ← Valley networks → | ?

← Period of warm, moist conditions | ?

4.5 ████ 4.0 ████ 3.5 ████ 3.0 ████ 2.5

← CO₂, N₂ atmosph

EARTH

Meteor bombardment | ← Definite evidence for life →

| ← Oldest rocks

earliest life, we can confidently say that it took no longer than about half a billion years. But it could have been much shorter, conceivably only millions of years.

So life on Earth seems to have arisen at the same time that water, the quintessential requirement for life, flowed in the valley networks on the surface of Mars. That's why planetary scientists think life might also have developed on Mars and why they want to search for fossil evidence of this life.

Life under a Greenhouse?

If Mars had a warmer, more clement climate in the past, what happened to it? Current theories suggest that

Mars started off with a thick atmosphere of carbon dioxide (as thick as or even thicker than Earth's present atmosphere) that outgassed from the planet's interior. This atmosphere acted much like a greenhouse by trapping infrared radiation, which kept the surface warm and allowed liquid water to flow.

But when the atmospheric carbon dioxide combined with liquid water, it formed a weak acid, carbonic acid. This reacted with rocks on the surface, transforming the carbon into carbonates such as limestone and dolomite. These carbonates then precipitated on the floors of oceans and lake basins. Eventually all the carbon dioxide in Mars' atmosphere

Loss of atmosphere ----------→

Outflow channels ----------- ?

Time, billions of years ago

2.0 ━━━━━━ 1.5 ━━━━━ 1.0 ━━━━━ 0.5 ━━━━━ Now

Buildup of atmospheric O_2 ----→

Microbial life ━━━ Higher life forms ━━━ Dinosaur extinction ━→

Vascular plants

ASTRONOMY: Lee Vande Visse

was exhausted, and the planet's surface became the cold, dry place we see today.

The same process would have happened on the early Earth (which we also think was kept warm by a thick atmosphere of carbon dioxide) except that on Earth, plate tectonics helped recycle the carbon dioxide. When two plates collide, one dives beneath the other in a process called subduction. Subduction carries the carbonate deposits to great depths, where heat decomposes the carbonates back to carbon dioxide gas, which then comes bubbling up in the volcanos created at the subduction zones.

Mars lacks plate tectonics, presumably because it is

smaller and therefore radiated away its internal heat faster than it could generate it. With no long-term mechanism to recycle the carbon dioxide back into the atmosphere, the climate deteriorated and became uninhabitable. (Some recent studies, however, suggest that Mars may have regenerated a thick atmosphere when hot lava flowed over the carbonate deposits and released the trapped carbon dioxide. The thicker atmosphere could have ushered in a return to relatively warm, wet conditions for a period of a few hundred million years.)

Now the key question arises, How long did Mars have liquid-water habitats suitable for life? The answer

to this question has two parts: How fast did Mars lose its carbon dioxide atmosphere, and how cold can it get before liquid-water habitats suitable for life disappear?

We now think that the initially thick atmosphere of carbon dioxide on Mars disappeared rapidly during the warm, moist conditions that existed prior to and perhaps after the end of the heavy bombardment about 3.8 billion years ago. But the rapid removal of the atmosphere lasted only as long as there were bodies of liquid water to absorb the carbon dioxide. As the atmosphere thinned, the greenhouse effect abated and temperatures eventually fell below freezing. From that point on, the carbon dioxide would have been removed ever more slowly and further drops in temperature also would have proceeded more slowly. This leads to the second part of the question, life's ability to survive increasingly frigid conditions. Ecosystems found in ice-covered lakes in the dry valleys of Antarctica provide a clue.

A Bit of Mars on Earth

These Antarctic valleys are extremely cold and dry. The average temperature is a chilly −20° Celsius and the net precipitation averages less than two centimeters per year. It never rains, and during most years the valley floors receive just a light dusting of snow. Because they are so cold and dry, these valleys are considered the most Mars-like places on Earth. In fact, they are the

most Mars-like places anywhere in the solar system — outside Mars, of course. At first glance, the dry valleys appear profoundly lifeless. And in fact their soils are so bereft of life that in the early 1970s they were used as analogs to the Martian soil during planning for the Viking biology experiments. Yet surprisingly, two major ecosystems exist within this barren, frigid desert.

In one, algae dwell under the perennial ice cover of the lakes on the valley floors. Despite ice thicknesses of three to six meters, enough sunlight passes through so the algae can photosynthesize. Nutrients and dissolved air flow into the lakes with water from the nearby glaciers that melt on the few warm summer days. In the other ecosystem, lichen lie just below the surface of certain sandstone rocks on the mountain slopes overlooking the valley floors. The rocks warm to well above freezing in the summer sunlight, retain moisture generated during the rare snowfalls, and shield the lichen against wind and ultraviolet radiation. Both ecosystems thrive during the few days each summer that liquid water is present.

These Earthly ecosystems prove that even if the average annual temperature is −20° Celsius, microbial life can prosper as long as even transient sources of liquid water exist. So it's not the average temperature that matters: It's the ability of Mars to create meltwater during the heat of summer. Using the Antarctic systems

COLD, DRY, AND INHOSPITABLE — that's Mars now, but there could still be a record of life if primitive organisms formed there billions of years ago.

Paul Hudson

existence. When something dies in a lake, its body can settle to the bottom and be preserved in the sediments, awaiting exhumation by scientists.

Large lake beds have the added benefit of being relatively easy for spacecraft to land in. The late Harold Masursky of the U. S. Geological Survey was the first to suggest that the deposits seen in some of the canyons of the Valles Marineris could be prime sites for searching for Martian fossils.

The Viking results strongly suggest that no life exists on the surface of Mars today; it is too dry, too cold, and too oxidizing. But the Viking landers only scratched the surface — both literally and figuratively — in their search for life on Mars. Underneath the surface it may be possible for liquid water to exist and persist for long periods of time. After all, the outflow channels offer direct evidence that liquid water has been present in subsurface aquifers on Mars throughout most of its history.

Lifeforms in such an aquifer could not obtain their energy from sunlight as photosynthetic organisms do on Earth. Instead, they would have to utilize chemical energy. Such chemosynthetic organisms appear on Earth in the form of methanogens, which consume molecular hydrogen and carbon dioxide and produce methane as a waste product. So a habitat for life on Mars could be based on a geothermal source of hydrogen together with atmospheric carbon dioxide and subsurface liquid water. We don't have any direct evidence that such a salubrious underground niche exists, but we can't rule it out either. The high-resolution camera and the thermal emission spectrometer on Mars Observer may help locate such geothermal hot spots — Martian Yellowstone Parks — if they exist.

To get to the evidence for life on Mars we will have to dig deep into the surface. This is true whether we search for fossil evidence of past life or try to reach present forms in their subsurface habitats.

The fascination with the search for life on Mars, both dead and alive, reflects humankind's overall quest for life beyond Earth. While our theories suggest that life should be widespread in the universe, we have no hard data to support this. We've detected no alien signals and found no oxygen-rich planets around distant stars. In our own solar system, Earth appears to be the only planet with an active biosphere. In fact, the evidence for liquid water on Mars, albeit billions of years ago, is the best clue we have to life elsewhere, and the search for life there is the first real test of our theories of how life began.

Earth and Mars led remarkably similar lives during the first billion years of their existence, and everything we know about how life originated on Earth hints that life likely arose on Mars too. Whether we find life on Mars or not, the search for it promises to teach us more about our origins and the origin of life in the universe. □

as a guide, my colleagues and I have estimated that the duration of liquid habitats on Mars, those able to sustain life similar to that seen in the Antarctic dry valleys, was from 500 million to a billion years. Although we don't know precisely how long it took life to evolve on Earth, we think it was much less than a billion years.

Just maybe life evolved on Mars too. And if it did, there's a good chance that evidence for this early life may still exist, because over two-thirds of the Martian surface is older than 3.5 billion years. The record of early Mars, unlike that of Earth, is still preserved, waiting to be examined. It's ironic that a planet like Earth has destroyed the past record of life as it has maintained that life. Although the Martian surface soil appears to have been altered by reactive chemicals, this should not have destroyed organic or fossil evidence for life deeper down. Thus, while Mars probably has no life on it today, it could hold the best evidence we have as to how life begins in an Earth-like environment.

Hunting for Martian Fossils

Where would we look for fossils on Mars? The best locations would be places where there had been a standing body of water, such as a lake. A lake is not only a good place to live when conditions get cold, as is shown by the Antarctic dry valleys, but also a good place to die if you want to leave a record of your

Christopher P. McKay is a planetary astronomer in the Space Sciences Division at NASA's Ames Research Center in Moffett Field, California.

ALL IN THE FAMILY

When asteroids collide, pieces fly. Tracking the fragments is giving astronomers new insights into the history of the asteroid belt.

by Dan Durda

Ever see a home video in which somebody drops something breakable? Wasn't it fun to run it in reverse and see all the scattered pieces come flying together again? With the help of large computers and sophisticated software, some astronomers are doing much the same thing — only it's not eggs or pumpkins they're putting back together but asteroids, giant rocks that orbit the Sun between Mars and Jupiter.

In this region, the asteroid belt, drifts a wasteland of rocky debris. The largest asteroid, Ceres, is only 1,000 kilometers in diameter — barely big enough to show a disk in a large telescope. All other asteroids are smaller, showing as pinpoints or streaks on long-exposure photos. These flying mountains slowly turn and tumble as each traces its lonely orbital path.

Because planetary scientists can't image the surface of an asteroid (except with expensive spacecraft), it seems hopeless to try to unfold the object's history or say anything meaningful about its geology. Yet astronomers have found that some asteroids — the broken fragments of larger ones — do hint at their backgrounds. When identified, these fragments offer a rare glimpse into the interior of asteroids. They also give astronomers a chance to learn more about the properties of their parent asteroids and the mechanism of their destruction.

The broken-asteroid story began 75 years ago, when Japanese astronomer Kiyotsugu Hirayama noted that a handful of asteroids share similar orbital properties. The members of these groups not only orbit at nearly the same distance from the Sun but also share similar orbital inclinations and eccentricities. Suspecting the asteroids within a group were somehow related to each other, Hirayama called the associations "families."

He wrote that explaining an asteroid family seems easy if you suppose that a large asteroid broke into a number of fragments, probably in a devastating collision with another large asteroid. The fragments — the children of the two parent bodies and now themselves individual asteroids — will follow their own orbits around the Sun. But as they do, each fragment "remembers" the orbit of its parent.

Cosmic Rock Crusher

Contrary to popular notions, the asteroid belt is a pretty empty place and collisions between major asteroids are rare. But they *do* happen . . .

Imagine a large asteroid, perhaps a couple hundred kilometers across, orbiting the Sun between Mars and Jupiter. For most of its life the asteroid has seen little more than a few crater-making impacts. Many more sand-grain-sized meteoroids drift through the belt than boulders, so most impacts have been small in scale, gardening the surface soil to a depth of several inches.

Now, however, imagine that a much larger object is approaching. Closing at more than 5 km a second, the oncoming chunk of rock — perhaps only a tenth the size of its target — slams into the asteroid.

Debris from the shattered projectile and the target asteroid is blasted away from the point of impact as an immense shock wave shudders through the asteroid. Unable to withstand the strain, the target asteroid fractures into many large pieces and countless bits of smaller debris. Fragments ranging in size from kilometers across to motes of dust promptly depart the scene of the collision at speeds up to hundreds of meters per second.

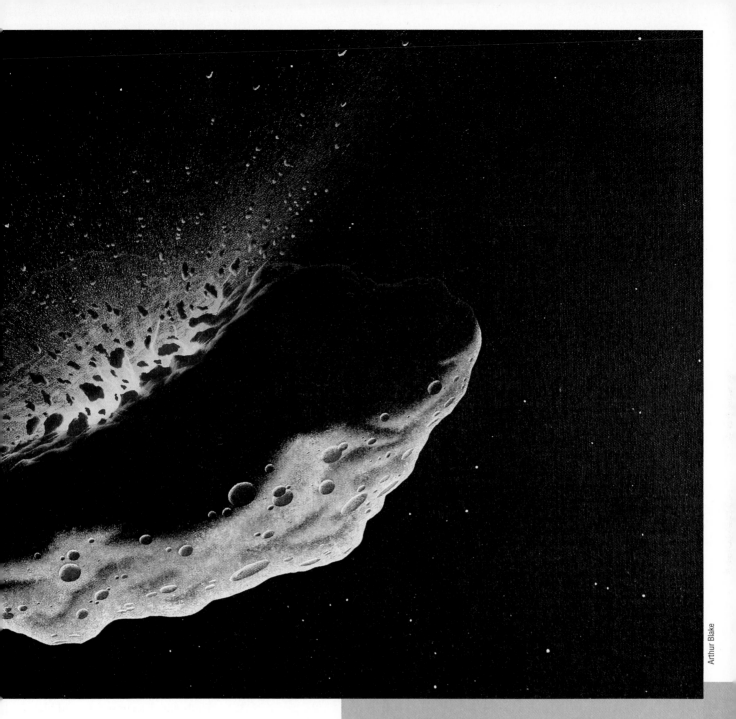

THE JARRING SLAM as two minor planets collide gives birth to an asteroid "family" with fragments sailing off on their own orbits.

Such speeds are small, though, compared to the asteroids' orbital speeds around the Sun. This means that while the newly formed fragments assume their own orbits in the asteroid belt, the new orbits are but variations on that of the parent. Thus an asteroid family is born.

Needles in Haystacks

Rounding up family members today, however, is a big challenge because they no longer orbit the Sun in tight groups. At the beginning, each family is a little clump of debris, but this stage soon passes. After only a few hundred years, their relative speeds will spread the members of a family into a ring of sorts around the original orbit (see pages 38-39). Then over a period of a million years or so, gravitational tugs from the planets, especially Jupiter, will completely smear the orientations of the orbits around the belt. At any given moment today, individual family members may lie scattered on opposite sides of the asteroid belt.

When this happens astronomers can recognize a family only after they carefully compare the orbital elements of lots of asteroids, most of which are probably unrelated in reality. Orbital elements are the mathematical quantities that define the shape and orientation in space of an orbit. Some orientation elements can be altered by the gravity of other planets, but orbital shape is more durable. Astronomers therefore plot invariable elements against each other, looking for similarities. When they do this, they find that members of a family tend to cluster on a graph (see page 39).

Hirayama found the three most prominent families

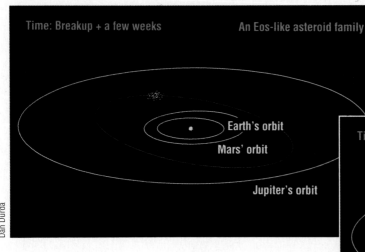

Time: Breakup + a few weeks   An Eos-like asteroid family

Earth's orbit

Mars' orbit

Jupiter's orbit

Dan Durda

LEAVING THE NEST

AFTER THE BREAK-UP, asteroid family members scatter throughout the main belt. Yet each retains an orbit that resembles in size and shape the orbit of its parent body.

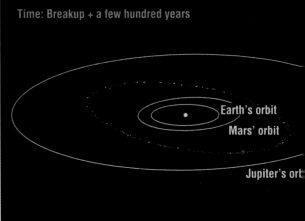

Time: Breakup + a few hundred years

Earth's orbit

Mars' orbit

Jupiter's orb

by eye. These were the Koronis, Eos, and Themis families, each named for its largest member asteroid. In Hirayama's time only a few hundred asteroids were known; today the number runs well beyond 5,000. As a result astronomers have found many new families, using means that range from eyeball searches to more objective mathematical methods. Unfortunately, some researchers claim to be able to identify more than a hundred groupings, while others consider that only the few largest families are real.

The disagreements arise from several causes, such as different astronomers' using different starting sets of asteroids. Early family classifications included fewer asteroids; as more asteroids become known, investigators can identify the less well-populated families that went undetected previously. Second, different ways of calculating the orbital elements yield different groups. And third, different statistical ways of distinguishing family groups from the background asteroids of the main belt can produce different families also.

Probably the most extensive list of asteroid families is that of Jim Williams of the Jet Propulsion Laboratory. In his latest work, he has identified 104 families, many quite small, by examining stereoscopic plots of the asteroids' elements (see page 40). He then puts those clusters he has visually identified through a statistical test to see if they really do stand out against the background of the other asteroids.

Another list of families comes from the work of Vincenzo Zappalà and his co-workers in Italy. They use a large set of asteroids, a refined theory for deriving elements, and an objective method of looking for non-random groupings of asteroids. Zappalà and colleagues have found 21 families that they consider to be statistically reliable, most of which match fairly well those found by other astronomers.

Putting Humpty Together Again

Finding families is only half the battle. If families are to help scientists understand the physical nature

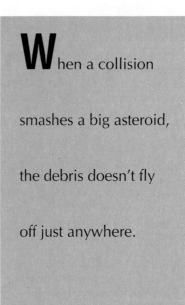

When a collision

smashes a big asteroid,

the debris doesn't fly

off just anywhere.

of asteroids, then astronomers must be careful that their lists contain families that are truly the results of catastrophic collisions and not chance groupings of unrelated objects. Clearly, since there is no unanimous agreement on the number of families or on which asteroids should be included in families, it's natural to ask if *any* of the families identified by various researchers are real.

Some asteroid scientists, particularly Jeff Bell of the University of Hawaii and Clark Chapman of the Planetary Science Institute, have attempted to answer this by looking to see if the families make geological sense. They reason that if a family originates in the destruction of a single parent asteroid, scientists should be able to reassemble the various members into a geologically believable parent body. Bell and Chapman use asteroid-type classifications that are based on how asteroids appear as viewed from Earth with spectroscopes and photometers. These classifications imply certain mineralogies that can be correlated with how rocks are known to form. Let's look at some families in more detail and see how the Williams and Zappalà lists of families compare when examined in this manner.

Hirayama found the three largest families in his pioneering study in 1918. The Koronis, Eos, and Themis families together comprise the majority of family asteroids. Of the 5,000 or so numbered asteroids, more than 860 belong to families and almost 600 belong to these three large families. Not unexpectedly, the three families are virtually identical in both the

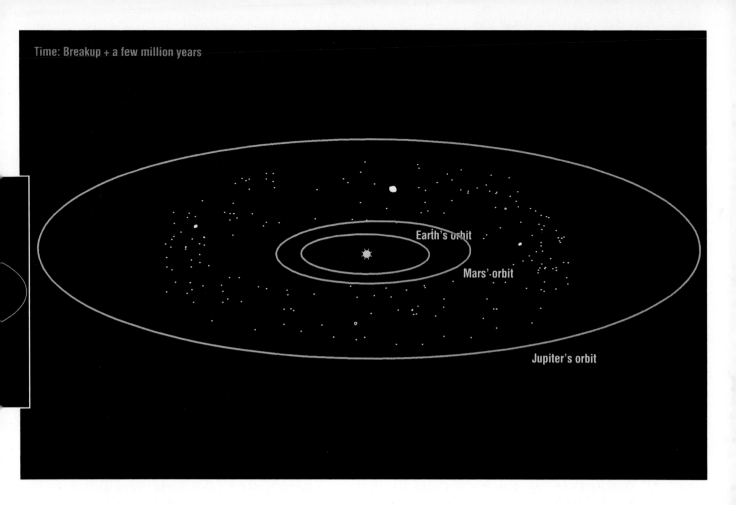

Earth's orbit

Mars' orbit

Jupiter's orbit

Williams and the Zappalà classifications. Moreover, all three are geologically consistent with the idea of a single-parent body since the asteroids within each family exhibit the same classification types.

The Koronis family resulted from the destruction of a modest-sized asteroid. Koronis itself is about 40 km across and some other members of the family are almost as large. This suggests the parent body broke apart completely in the collision. Observed from Earth, the Koronis family members show large changes in brightness, indicating that many are tumbling fragments with irregular shapes. Astronomer Rick Binzel of MIT studied their rotation rates and concluded that the Koronis family may be relatively young. Its members appear not to have undergone many collisions, which would alter their shapes and spins.

Interestingly, astronomers may even be able to say something about where in its orbit the precursor asteroid was disrupted. Koronis members

show very little spread in their inclinations, which hints that the parent broke apart near the high or low point in its orbital excursion from the plane of the solar system.

When astronomers plot the Eos family's orbital elements, those of the larger members all fall in a distinct core with the smaller members lying at its edges. The spread of elements within a family tells about the range of velocities with which the members were ejected when the parent fragmented. The core of a family thus represents the slow-moving, larger pieces, while the

GATHERING THE CLANS

FAMILY REUNIONS OCCUR when astronomers plot one orbital property against another. The similarities in their orbits bring family members (shown in color) together on the graph.

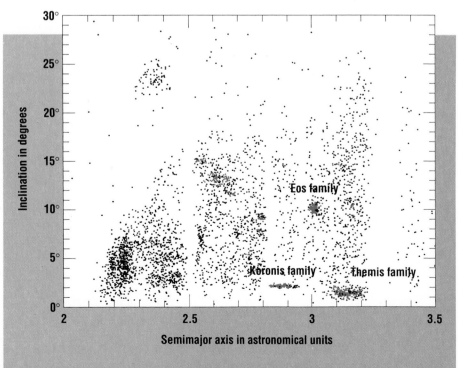

Eos family

Koronis family

Themis family

Inclination in degrees

Semimajor axis in astronomical units

Dan Durda

BIG FAMILY OUTINGS

TO CHECK FOR FAMILIES, astronomers sometimes use 3-dimensional graphs to seek out family members. This stereo pair has semimajor axis as the dimension coming out of the page toward you.

(To see the effect without a stereo viewer, hold a piece of cardboard upright between the pair of images and illuminate both sides equally. Look down with one eye on each side of the cardboard, and relax your eyes until the two images fuse into one. Family member asteroids should appear to clump above the page.)

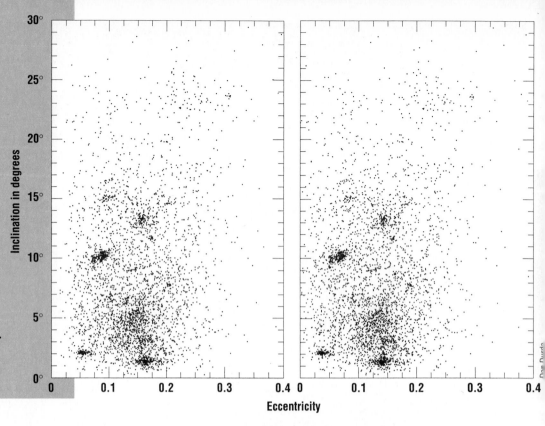

outliers represent smaller debris ejected from the scene at higher speeds.

Like the asteroids of the Koronis family, Eos members belong to the same type and many have properties that are unique to them. The fact that the members of this family display a distinct spectral class says not only that the parent asteroid was rather homogeneous, but also that the family truly resulted from the fragmentation of a single asteroid and is not a cluster of unrelated objects.

The plot of the very large Themis family also shows a compact core surrounded by less-dense clumps. Most of the non-core members of the Themis family lie to one side of the core, which may indicate that these fragments came from one side of the parent body. Williams estimates that the parent object lost some 60 percent of its mass in the collision and that Themis itself, some 227 km across, is the intact remnant of the parent. In what might be considered an enormous cratering event, much of one hemisphere of the Themis parent was blasted into space. The remaining hemisphere blocked the debris from traveling into the other side of the sky. The less-dense clumps in the outer parts of the family may represent subsequent collisions that further broke apart some of the resulting family members. Astronomers suspect the family trees of many asteroids may show these hierarchies of fragments.

A similar event but one much smaller in scale seems to have created a tiny family found by both Williams and Zappalà. It is associated with the asteroid Vesta. This is

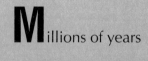

Millions of years after the fact, astronomers can still spot the runaway pieces of an asteroid's break-up.

extremely interesting since Vesta is well known for having unique spectral properties that indicate it has a basaltic crust, unlike other asteroids. However, Vesta's spectrum changes as it rotates. Vesta seems to have a hole in its crust, revealing olivine-rich rock in the lower crust and mantle below. Binzel reported that the sparse family of little asteroids associated with Vesta shares its spectrum and probably represents blocks of crustal rock excavated from the surface long ago. When a spacecraft eventually visits Vesta, astronomers expect to see at least one very large impact scar on its cratered surface.

Merged and Blended Families?

The Zappalà and Williams lists of the big families like Koronis, Eos, and Themis largely agree, but there can be quite a bit of disagreement on smaller families.

Asteroids Nysa and Hertha have been linked in many previous family classifications. They appeared to be pieces from the destruction of a parent asteroid that melted throughout and differentiated — that is, it became stratified, with its heavy material sinking to make a core. Nysa has a diameter of about 73 km, while Hertha is 82 km across. Nysa's spectrum links it to a type of stony meteorite that astronomers think comes from the mantles of bodies that are differentiated. Hertha could be the metallic core (or part of it) of the parent asteroid.

Most recent family classifications, however, do not include a well-defined Nysa-Hertha family. For instance, neither asteroid appears among the statistically significant families of Zappalà. Williams

GASPRA COMES FROM A GROUP of minor planets that may be associated with the asteroid Flora. Judging by the extent craters have marked its surface, astronomers think Gaspra may have been a separate body for the last 200 million years.

JPL/NASA

places them into two separate, but not widely separated, smaller families. There are also a number of small, rare, and primitive asteroids associated with the Nysa group, which would be difficult to understand from a mineralogical standpoint. Jeff Bell and others have even suggested that Nysa and Hertha are interlopers in a more homogeneous family composed of primitive asteroids.

Bell examined both the Zappalà and Williams lists of asteroids and found that most of the geologically questionable Williams families do not correspond to any of the Zappalà families. Although it seems that many of the Williams families cannot derive from a single parent asteroid, most of the Zappalà families can. Bell concludes that in general the Zappalà families are closer to reality than the families identified by Williams.

This is not to say that all the small Williams families are false. Clark Chapman notes that some of these families are distinct in composition from the background asteroids around them. This suggests that the asteroids within the families may relate to one another, even though they may not seem to make mineralogic sense at first. Williams suggests that some of his families that contain a mix of types may be examples of families that suffered several collisions over the age of the solar system. Collisions with the parent body in its early days, when impacts were more frequent, would distribute primitive material, while later impacts would excavate more geologically processed material. Thus some families might preserve a record of collisions with an evolving asteroid, and these families would then contain what appear to be discordant types.

Williams cautions that it might be better to use families to explore relationships between types of asteroids rather than vice versa. It is also wise to remember that the mineralogic interpretations of some asteroid types may not necessarily be correct. For example, the nature of the very common stony, or S-type, asteroids is still a source of debate among researchers. Planetary scientists hope that data from the Galileo encounters with Gaspra and Ida, both S-type asteroids, may help. In fact, both Gaspra and Ida are members of families: Gaspra lies in a complex cluster perhaps associated with the asteroid Flora, and Ida is a member of the Koronis family.

Clearly, astronomers will have to study these clusters of asteroids more carefully before they can understand their origin. As more and fainter asteroids are discovered and studied, missing pieces to the family puzzles will fall into place. Astronomers expect to identify many smaller families in the outer asteroid belt. The entire asteroid population, in fact, will likely display a fractal nature: families within families within families.

When researchers have gleaned from observations the physical properties of more individual members of asteroid families, they will be better able to reconstruct the parent asteroids and the circumstances of their destruction. It's a little ironic that astronomers can best bring asteroid families back together again by exploring the details surrounding how they broke up! No astronomer anticipates running the video of the asteroid belt in reverse right back to the beginning. But tracking down asteroid families is proving to be a powerful way to unlock some closed portions of the solar system's history. □

Dan Durda is an astronomer at the University of Florida who studies the collisional evolution of asteroids. He is also investigating how much solar system dust comes from asteroid collisions.

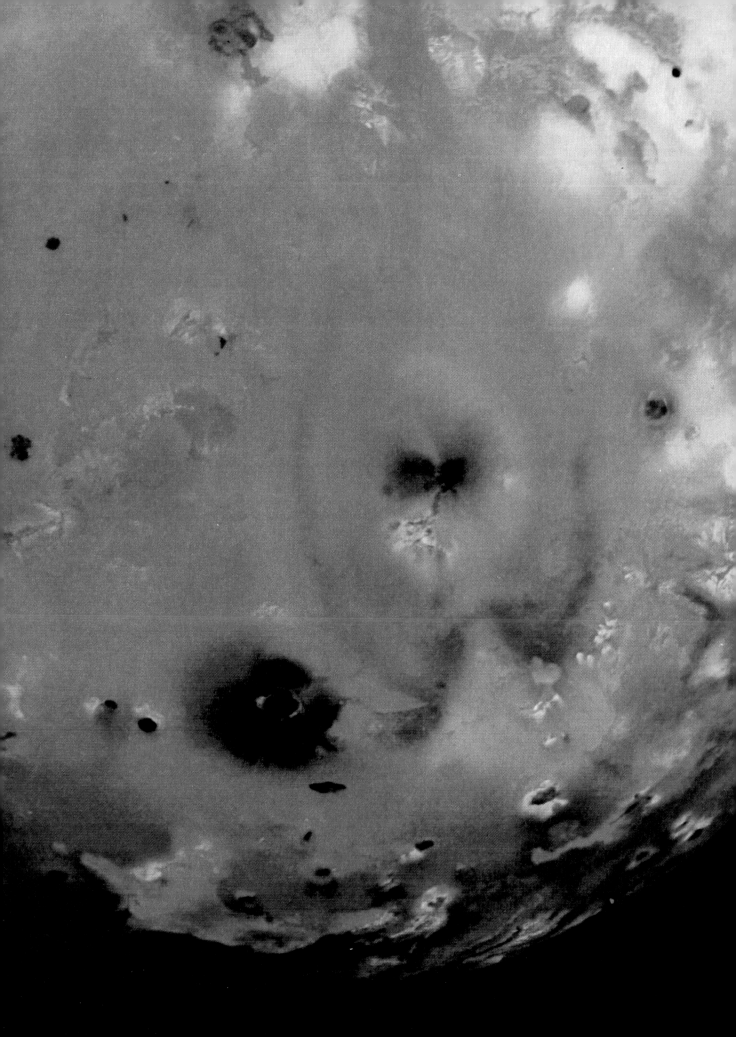

PLANETARY SATELLITES

The Violent Volcanos of

NASA/JPL

The volcanic fires of Jupiter's extraordinary moon Io rage at levels far higher than any seen elsewhere in our solar system.

by Richard Talcott

F ire and brimstone. The words bring to mind a hellish land of molten lava, of volcanos spewing sulfurous fumes high into a thin atmosphere. To astronomers, however, this place is not one to fear or avoid, but to scrutinize. For this environment belongs to a real world, Jupiter's remarkable satellite, Io.

When the Voyager 1 spacecraft discovered erupting volcanos on Io during its 1979 flyby, most astronomers were astonished by their sheer magnitude. Greeting Voyager's cameras were nine separate eruptions, each of which dwarfed anything ever seen on Earth. The images painted a picture of a world literally turning itself inside out.

Though spacecraft first revealed the volcanos' presence, astronomers are not sitting idly by waiting for the next space mission — the Galileo orbiter — to arrive at Jupiter in 1995. Thanks to sophisticated detectors on ground-based telescopes, they can now monitor Io from Earth. What they've found is a world that remains the most volcanic body in the solar system but one whose activity varies significantly over short periods.

Redder than Mars

Even before Voyager discovered active volcanos, Io appeared intriguing. Observers have studied this tiny world ever since 1610, when the Italian astronomer Galileo discovered Io and Jupiter's three other large satellites — Europa, Ganymede, and Callisto. For over 300 years, astronomers thought that the four Gali-

A HEART-SHAPED PATTERN OF DEBRIS surrounds Io's volcano Pele, shown here from directly above. Pele's sulfur-rich eruption was the largest seen by Voyager, spewing volcanic ash across more than 1,000 kilometers of Io's surface.

lean satellites were similar to each other and to Earth's own satellite, the Moon.

That view didn't change until this century, when detailed observations with better equipment revealed Io as an oddball. In the 1920s, observers noted that Io's brightness and color varied significantly as it orbited Jupiter. And overall Io turned out to be the reddest body in the solar system, redder even than the Red Planet, Mars.

Two events in 1964 added to the intrigue. First, observations seemed to show that Io brightened abnormally as it emerged from Jupiter's shadow after an eclipse. (Surprisingly, the question of whether this brightening is real has yet to be settled.) Second, powerful radio bursts from Jupiter correlated with Io's orbital position, showing that Io was somehow linked to the Jovian magnetic field.

In 1973 the Pioneer 10 spacecraft performed the first closeup reconnaissance of Jupiter. Though it did not return any photos of Io, Pioneer 10 did provide an accurate value for the satellite's mass. When combined with the known diameter of Io (3,630 km), the mass yielded an average density of 3.6 grams per cubic centimeter, some 7 percent greater than the Moon's and what would be expected for a body of pure rock.

Perhaps most puzzling of all, ground-based observations in the infrared showed that when Io passed into Jupiter's shadow, its temperature fell faster at some wavelengths than others. Io baffled astronomers in the 1970s because so many of the observations simply didn't make sense. But then astronomers were thinking that Io was a cold, geologically dead world.

All that changed in March 1979,

when Voyager 1 zipped through the Jovian system. Highly detailed photos of Io showed no impact craters on the satellite's surface. If Io were a dead world, it should have been pockmarked with impact craters, so their absence meant some process had to be destroying them.

Plumes in the Ionian Sky

A few days later, astronomers solved the mystery — they saw active volcanos on Io! No less than nine volcanos were spewing vast plumes of material tens to hundreds of kilometers above the Ionian surface. The intensity of the volcanism is so high that Io resurfaces itself at a rate as high as 10 cm per year.

Eruptions on Io fall into two major categories. The volcano Pele typifies the first kind and, in fact, is the only example of this type Voyager actually saw erupting. Pele-type volcanos produce umbrella-shaped plumes that reach heights of 300 km and leave deposits over an area 1,000 km or more across. To reach such heights, the volcanic material must be ejected at speeds of about 1 kilometer per second (km/s).

Pele-type eruptions appear to be driven by sulfur vapor and are short-lived, lasting probably only several days to a few weeks. Pele itself had shut down by the time Voyager 2 flew by four months after Voyager 1. Two other volcanos, however, Surt and Aten, showed fresh deposits that indicated they experienced Pele-type eruptions during the interim.

The other eight plumes, seen by both Voyager spacecraft, are all similar to the volcano Prometheus.

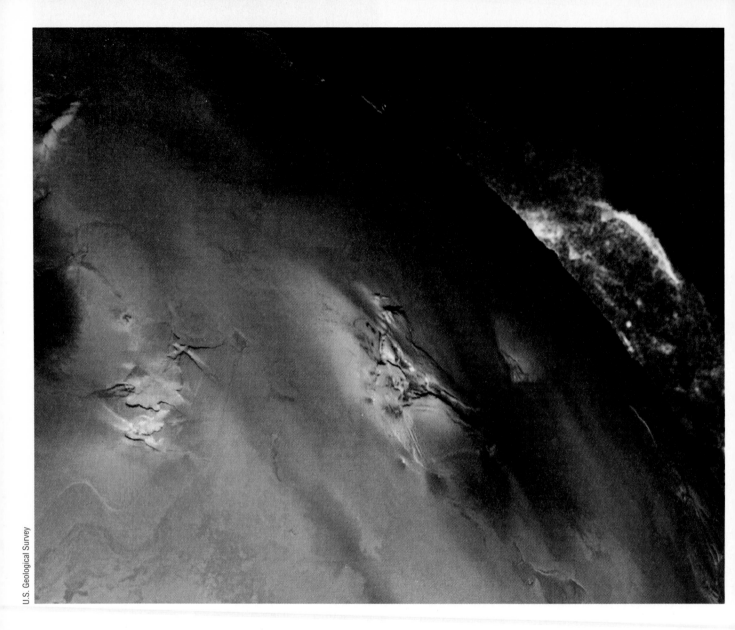

U.S. Geological Survey

PELE ERUPTS, shooting a volcanic plume some 300 kilometers into Io's thin atmosphere. The eruption was brief, however, having ceased by the time Voyager 2 flew by four months after Voyager 1 took this image.

Prometheus-type eruptions are smaller than that of Pele, extending from 50 to 200 km above Io's surface and depositing material across a few hundred km. They spew out ash particles at 0.5 km/s and apparently last for years at a time. The best guess is that these eruptions originate in reservoirs of liquid sulfur dioxide at temperatures of about 400 K.

One volcano deserves special mention. Loki, though classified as a Prometheus-type volcano, may actually be an intermediate type. For a few days during the Voyager 2 encounter, one of its two plumes increased to near Pele-like levels.

Voyager 1 also mapped the temperature of Io's surface with its Infrared Interferometer Spectrometer. (Voyager 2 passed too far away to measure heat from Io.) The instrument found hotspots associated with some volcanic plumes and with some other volcanic features that were not erupting. The temperatures of these hotspots ranged from nearly 400 K for Prometheus-type volcanos to 600 to 700 K for Pele.

The Tide Is High

What's the ultimate source of the heat needed to generate all this volcanic activity? Astronomers generally consider two methods as ways for planets or satellites to produce enough internal heat to drive surface geological activity. The first is by the decay of radioactive elements. Long-lived radioactive elements such as uranium, thorium, and potassium can generate heat for billions of years as they decay. That's what powers volcanos and plate tectonics on Earth.

However, the smaller an object is, the less radioactive material it contains and the faster it radiates heat into space. Molten lava once flowed on the Moon's surface, for example, but that was billions of years ago. The Moon has now cooled to the point where its solid crust is several hundred kilometers thick. Because Io has roughly the same size and mass as the Moon, radioactive heating should play little or no role in the ongoing volcanism there.

The second method of internal heating — tidal heating induced by a neighboring body — is the one at work in Io. This theory was first proposed for Io by Stanton Peale of the University of California at Santa Barbara and his colleagues and published just three days before the Voyager 1 encounter with Jupiter. Jupiter's gravity raises a tidal bulge in the solid body of Io in the same way that the Moon produces ocean tides and a tidal bulge on Earth, and Earth

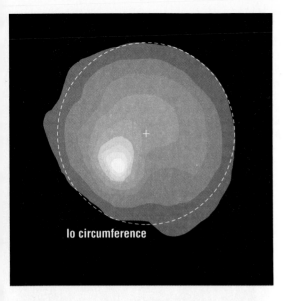

Io circumference

PELE RETURNED TO LIFE in March 1990 after a decade of apparent silence. Pele's reawakening shows up in this ground-based contour plot (top) and infrared image (above) as a bright "hotspot" in the lower left corner of Io.

produces a tidal bulge on the Moon. And as with the Moon, tidal forces caused Io long ago to fall into synchronous rotation about Jupiter. This means Io rotates once on its axis every time it revolves once around Jupiter and thus always keeps the same face pointing toward Jupiter.

These tidal forces would produce no heating if Io's orbit around Jupiter were a perfect circle. But Io follows a slightly elliptical path, and when it is closest to Jupiter it travels slightly faster than average, and when it is farthest from Jupiter it travels slightly slower. Io keeps spinning on its axis at the same rate regardless of where it lies in its orbit, forcing its tidal bulge to wobble back and forth. Jupiter's

gravity always works to pull Io's tidal bulge into a straight line with Jupiter. This causes constant flexing of Io's crust and generates Io's internal heat.

If Io and Jupiter existed alone in space, this action would eventually force Io's orbit into a perfect circle. But the inner three Galilean satellites orbit in resonance — each time Ganymede completes one orbit around Jupiter, Europa completes two and Io four. The periodic alignment keeps the elliptical orbit of Io from becoming circular and keeps the internal fires burning brightly.

Though tidal heating explains how Io generates so much volcanic activity, one troubling question remains. Detailed measurements made by Alfred McEwen of the U. S. Geological Survey and his colleagues show that the amount of heat flowing from Io during the Voyager 1 flyby is probably significantly greater than the average amount of tidal heating. Apparently Io either was anomalously active during the Voyager encounter or the amount of tidal heating varies with time.

Seeing with Infrared Eyes

Once the Voyager spacecraft left Jupiter behind, astronomers were forced to observe Io from Earth. But could they learn anything of interest about Io? After all, the most striking aspect of Io is its volcanic activity,

something most astronomers never even suspected from observations predating the Voyager flybys.

Two reasons explain why Earth-based studies now prove so useful in monitoring Io. First, astronomers know from the Voyager data that volcanos exist on Io and where they are located, which helps observers interpret what they see in the telescope. And second, infrared detectors have improved so much in the past decade that Earth-based telescopes can now resolve individual hotspots on Io's surface.

Infrared observations are the most useful because hotspots contribute most of the energy radiated by the satellite at infrared wavelengths. Though Io can be studied almost any time Jupiter is visible, the best time to observe it in the infrared is when it lies in Jupiter's shadow. During these eclipses, no sunlight reaches Io and thus no solar infrared radiation is reflected toward Earth to mask the volcanic hotspots. The main drawback to observing Io during eclipse is that its Jupiter-facing hemisphere always faces Earth at that time and the other side can't be seen.

Another good time to observe Io is while it is passing behind the limb of Jupiter. By measuring the brightness of the satellite over time, astronomers can discern when individual hotspots disappear behind Jupiter's limb. As with eclipses, however, these occultations permit researchers to view only the Jupiter-facing hemisphere. To see the other side of the satellite, they must observe when Io lies on the near side of Jupiter. A particularly good time to observe is when another Galilean satellite passes in front of Io,

VOLCANO ECLIPSE BEGINS

Europa

Pele

Loki

Io

VOLCANO ECLIPSE ENDS

ASTRONOMY: Thomas L. Hunt after P. Descamps

IO'S SISTER SATELLITE EUROPA passed in front of Io on February 20, 1991, slowly covering and uncovering hotspots associated with Io's volcanos (above). Carefully measuring Io's infrared brightness during this occultation revealed that the most active volcano at that time was Loki (below).

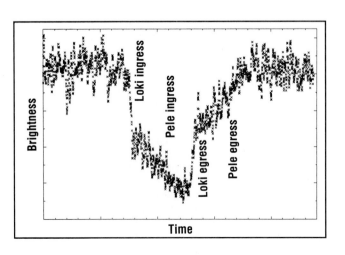

IO AND EUROPA HOVER ABOVE JUPITER — the trio of objects responsible for Io's volcanos. Tidal forces induced in Io by Jupiter stoke the satellite's internal fires, which are maintained because Europa keeps Io's orbit from becoming circular.

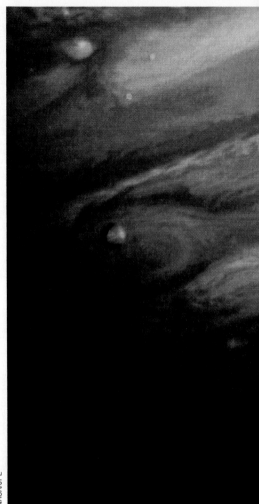

NASA/JPL

which happens several times during a roughly one-year period every six years (see "Watch Jupiter's Moons Play Tag" by John E. Westfall, January 1991).

Not only can infrared detectors resolve many of the hotspots detected by Voyager, they have even discovered a few not seen by Voyager. John Spencer of Lowell Observatory and his colleagues have been monitoring Io over the past several years using an infrared camera on the 3.2-meter NASA Infrared Telescope Facility on Mauna Kea in Hawaii. Observations they made in December 1989 when Jupiter eclipsed Io revealed a previously unknown hotspot. They dubbed the hotspot "Kanehekili" after a Hawaiian thunder god.

For the past several years, several groups (including Spencer's) have been following Loki, the largest hotspot on the Jupiter-facing hemisphere. Loki's brightness varies greatly — it was bright in late 1989 and early 1990, faint in spring and autumn 1990, bright again during the winter of 1990/1991, and faint between October 1991 and May 1992. Loki appeared 16 times brighter in December 1989 than it did in December 1991.

Loki appears significantly cooler when it is quiescent than when it is active. On January 24, 1991, during an active phase, Loki's temperature measured roughly 480 K. During a quiescent phase one year later, the hotspot's temperature dipped to about 355 K. The observations suggest that Loki's changes in brightness arise mostly from changes in temperature instead of changes in the area of the exposed volcanic materials.

The Return of Pele

Only one volcano out-erupted Loki at the time of the Voyager 1 flyby — Pele. But Pele went quiescent after that, escaping notice when Voyager 2 flew by just four months later and apparently remaining dormant for over a decade. Even detailed observations of a 1985 occultation of Io by Callisto showed no sign of activity. But Pele was far from extinct.

Brian McLeod and his colleagues

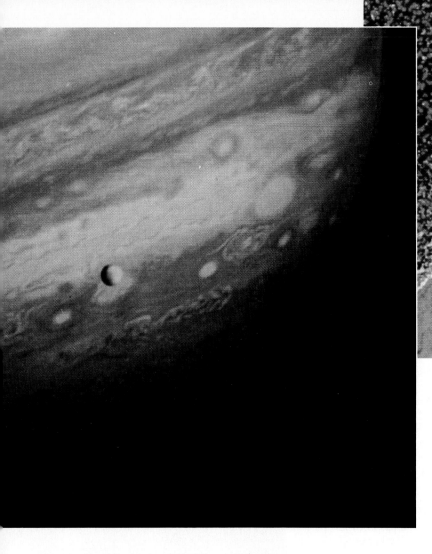

at the University of Arizona's Steward Observatory systematically observed Io with the Multiple Mirror Telescope during November 1989 and March 1990. Using a high-resolution imaging technique called speckle interferometry, they viewed four different central meridians of Io to gain a view of the entire object. And on March 8, 1990, while observing the hemisphere that faces away from Jupiter, they detected Pele (see page 43).

The observations imply that Pele can erupt at decade-long intervals, though the lack of continuous coverage makes it impossible to say how long an individual eruption may last. Pele is particularly hard to observe from the ground because it always faces away from Earth during the prime observing times when Jupiter eclipses or occults Io.

In fact, very few ground-based observations of this hemisphere have ever been made except during the series of mutual events among the Galilean satellites. The last such series occurred between late 1990 and early 1992, and several groups of astronomers were ready.

Jay Goguen of the Jet Propulsion Laboratory and his colleagues ob-
served Europa passing in front of Io eight times in early 1991. During a typical occultation, Europa travels about 10 km/s relative to Io. This means that measurements taken every few tenths of a second can resolve details as small as a few kilometers across on Io.

Goguen's team found both Loki and Pele active during that time and both varied significantly. Loki grew about 50 percent brighter during February alone, so astronomers got a bird's-eye view of the evolution of a major eruption. And on two occasions the researchers resolved Loki's "hotspot" into hotspots — two narrow fissures 20 to 30 km wide and about 100 km apart.

P. Descamps and his team from the Bureau des Longitudes in Paris, France, observed Europa occulting Io in February 1991. They also detected Loki and Pele, though Loki was much brighter. And Robert Howell and B. Uberuaga of the University of Wyoming observed several occultations, again showing both Loki and Pele, with Loki some ten times brighter. The mutual events confirmed that Pele was back, though at a much lower level than during the
Voyager 1 flyby, and continued to show Loki as a long-lived, highly variable hotspot.

Twenty Times Closer

Though ground-based observations will continue to play an important role, they can't come close to matching the detail seen from a spacecraft. The Galileo orbiter is now well on its way to Jupiter and will arrive in about two-and-a-half years. On December 7, 1995, the spacecraft and its cargo of sophisticated instruments will hurtle past Io at a distance of only 1,000 km. This is some 20 times closer than Voyager 1's flyby distance and, with Galileo's advanced camera, should yield photos with resolutions 50 times better (about 20 meters). Over the succeeding two years, Galileo will monitor Io's volcanos from a few hundred thousand kilometers away.

No one knows what Galileo will find. After all, the intensity of the volcanos varies markedly and quickly, so astronomers may see a whole different set of active volcanos. And the surface itself could look surprisingly different than the surface seen by Voyager. After all, every year the volcanos deposit about another 10 cm of material on the satellite's surface, so more than a meter of new material should cover the Voyager-era surface. Though the Voyager encounters provided unparalleled drama as volcanos erupted before the spacecraft's electronic eyes, Galileo promises to far surpass its predecessors and help astronomers understand the inner workings of the strange world of Io. □

Neptune's Weather Forecast
Cloudy, Windy and Cold

Why does a planet so far from the Sun have the fastest winds in the solar system? Planetary astronomers think they've found the answer deep within Neptune.

by Sanjay S. Limaye

Planets that have atmospheres have weather, and Neptune is no exception. In the almost two years since Voyager revealed Neptune's remarkable atmospheric features, planetary "meteorologists" like myself have been trying to understand this blue world's weather.

Neptune is a strange environment for weather: a place where, after encountering a bland Uranus, we expected to find little — and where we were surprised and rewarded by a richness of phenomena that have been challenging to understand. Voyager found Neptune to be a planet with active cloud forms and features reminiscent of those on Jupiter and even Earth. Neptune turned out to be the windiest planet in the entire solar system — with "breezes" reaching speeds of over 2,100 kilometers per hour. These findings were a surprise because Neptune's clouds and winds didn't fit what previous planetary encounters had led us to expect.

On Earth, solar heating drives the weather and winds (heat flowing up from the interior contributes little), and solar energy is abundant at Earth's distance from the Sun. But farther from the Sun, as solar energy dwindles, internal heat matters more. Voyager showed that Jupiter, which receives 1/27th Earth's sunlight but which has abundant internal heat, has a highly active atmosphere. Saturn, with 1/90th our solar energy and less internal heat, is quieter. Uranus, which gets 1/370th Earth's solar energy and has no detectable internal heat flow, displays a bland, deeply hazy atmosphere with little variation over the short or long term. And Neptune? It orbits where it receives only 1/900th the energy we do. With less than half the solar energy at Uranus and only a small internal heat source, Neptune could have been blander than bland.

Yet it wasn't. And over the months since the encounter, scientists have been studying the images — making movie loops to show Neptune's changing clouds, measuring their motions, and theorizing about their origin and the processes that drive them.

Clouds of Neptune

Voyager's biggest surprise was that Neptune even *has* clouds and storms. The first atmospheric feature seen in early Voyager images turned out to be one of the largest cloud systems not just on Neptune but in the whole solar system! Soon, by analogy with Jupiter's Great Red

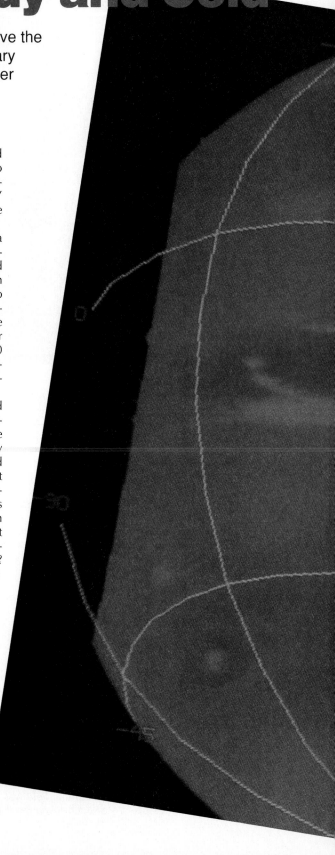

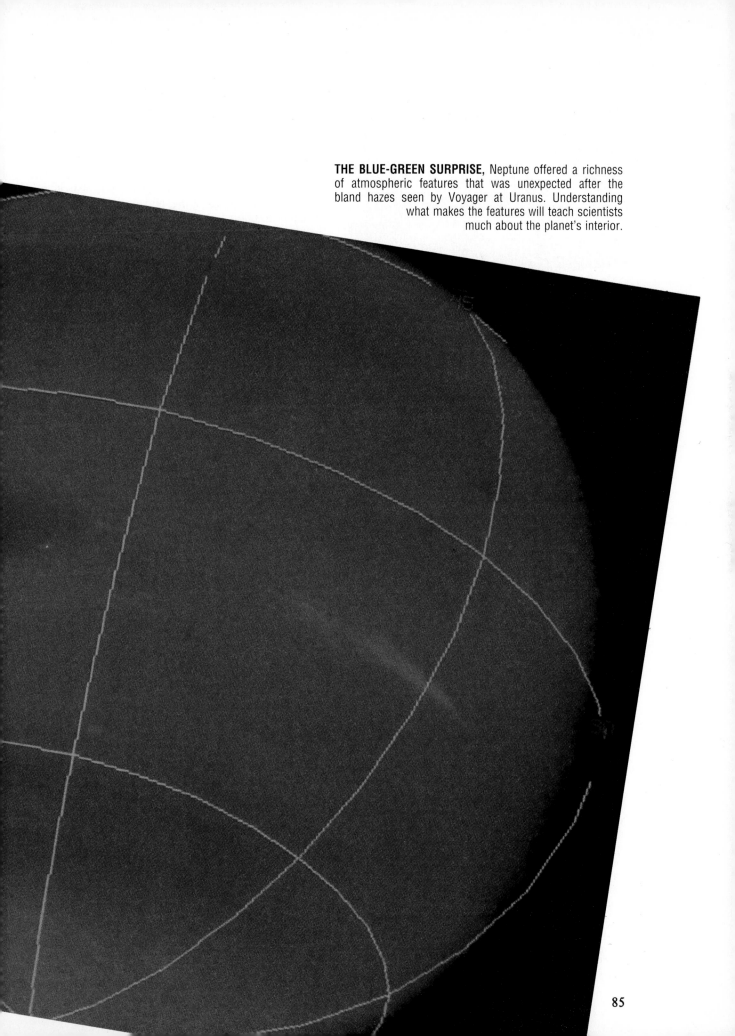

THE BLUE-GREEN SURPRISE, Neptune offered a richness of atmospheric features that was unexpected after the bland hazes seen by Voyager at Uranus. Understanding what makes the features will teach scientists much about the planet's interior.

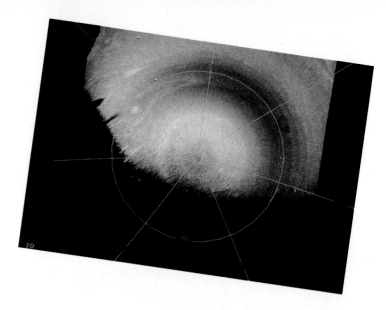

cloud appears tied to the global circulation, as are Earth's equatorial clouds.

At 30° north, Voyager saw white clouds that look like jet contrails — long, narrow, and straight. Such streamer clouds straddle the Dark Spot near its mid-section in one or two strings and also occur at 30° north. High-resolution images of the clouds show that they cast shadows on the main cloud deck below. From these, Voyager scientists estimate that the clouds drift 75 to 100 kilometers above the main cloud deck. At mid-southern latitudes there occur similar but much shorter clouds.

The Great Dark Spot has a feature called the southern companion, which appears linked to an intriguing kind of rippling wave cloud. On Earth these occur where a mountain range smoothly lifts a jet of air and gives its downstream flow a gentle ripple. But on Neptune what provides the uplift? One possibility is that the Dark Spot is a column of warm rising gas that forms a kind of "bump" in the Neptunian atmosphere.

Here Today, Gone Tomorrow?

Another clue to the source of energy behind the Neptunian cloud features is their lifetime. In planetary meteorology, size generally correlates with longevity. Jupiter's enormous Red Spot has been seen telescopically for 300 years at least. Elsewhere on Jupiter (and on Saturn) clouds smaller than Neptune's Dark Spot persist for at least several planet rotations. But Neptune is quite different: with the exception of the Great Dark Spot and DS2, its clouds evolve very rapidly. For example, features around the south pole cannot be followed for more than a few hours, let alone one Neptunian day. But what makes the clouds form and disappear so quickly is unknown.

Looking at larger-scale features, Neptune joins Jupiter and Saturn in having a strongly banded atmosphere, which sets them apart from Uranus. Scientists believe the rapid rotation of these planets and the relatively small differences in temperature from the equator to the poles are key reasons for the bands to exist. Earth and Mars, by contrast, have relatively large equator-to-pole temperature differences and long rotation periods; their thin atmospheres have turbulent, somewhat disorganized banding. But what controls the number of bands and widths? These are basic questions that have no answers as yet.

Winds of Neptune

Whatever the origins of the cloud features, tracking the velocity of individual clouds provides the easiest way to measure Neptune's winds. The method assumes, however, that small clouds will flow with the prevailing winds and indicate both direction and speed. Unfortunately, how much a given cloud shares in the local wind speed is not known very well. (In hurricanes, for example, clouds in the spiral bands may exceed 100 miles per hour but the storm as a whole may crawl at only 20 mph.) Used with caution, however, cloud motions determined from spacecraft images remain useful in mapping global winds.

Spot, this feature in Neptune's southern hemisphere was dubbed the Great Dark Spot. Comparable in size to the Red Spot, Neptune's Dark Spot is even centered at roughly the same latitude as the Red Spot is on Jupiter. And like the Red Spot, it appears to be rotating counter-clockwise, just as any high-pressure system in Earth's southern hemisphere does. (These similarities to the Red Spot give us hope we'll eventually understand its existence.)

But the Dark Spot has peculiarities too. First, it appears to vary more in size than the Red Spot. Second, Voyager movies show that while the Dark Spot retains an elliptical shape, it oscillates oddly. When the Dark Spot's north-south extent is shortest, the Spot stretches farthest east-west, and vice-versa. Third, the Red Spot is bounded by an easterly jet to the north and a westerly jet to the south, so it "rolls" between two opposing flows of wind like a ball bearing. But the Dark Spot lies wholly within a rapid westward flow and small clouds appear to move over it in some images. Finally, just as the origins of the Red Spot are unknown, those of the Dark Spot are equally mysterious, although it has probably lasted a few years at least and is very likely a semipermanent feature.

Some clues to the origins of Neptune's features may lie in its strong preference for particular types of cloud at particular latitudes. For instance, at the south pole Voyager spotted a tiny bright speck of cloud lying almost exactly at 90° south. Surrounded by bands of dark and light haze, this feature is unique to Neptune.

At 54° south, Neptune has a second dark spot called the Dark Spot 2, or just DS2. Similar in appearance to the Great Dark Spot and also rotating counter-clockwise, DS2 obeys one rule violated by its larger relative: DS2's center lies in a latitude where the winds to its south blow east while those to the north blow west. However, DS2 appears to be an obstacle in an otherwise uniform river of wind: the flow is gently deflected around the spot, much like a small pebble wrapped in a big spool of ribbon.

On Earth the equator is loosely marked by a band of cumulus clouds paralleling the equator (although not always straddling it). On Neptune we see a similar kind of cloud right on the equator, particularly in images made in the methane band. This methane-rich

NEPTUNIAN WIND SPEEDS

Latitude

40°
30°
20°
10°
0°
-10°
-20°
-30°
-40°
-50°
-60°
-70°
-80°
-90°

-1956 -1630 -1304 -652 -326 0 326 652

Velocity (kilometers per hour)

Other methods can yield wind speeds indirectly and let us check the speeds given by cloud motions. Two such Voyager experiments were the infrared interferometer and spectrometer (IRIS) and the radio occultation experiment. The IRIS results told us about winds by means of a meteorological relationship between the horizontal variation of temperature and variations of wind with altitude. The radio occultation experiment obtained wind speeds indirectly by the variable bending of the radio beam as it traversed the atmosphere on its way to Earth.

Of course, the idea of wind carries with it the idea that it is moving relative to something. On Earth we measure winds relative to the surface — that is, using a reference frame that rotates with the solid Earth. For Neptune or any other gaseous planet, planetary meteorologists have chosen as a reference frame the time the deep interior takes to make one complete rotation. Since this region is hidden from view, we measure this "day" by the rotation rate of the planet's radio emission.

Long before Voyager, scientists suspected that Neptune had winds. Michael Belton and his associates at Kitt Peak National Observatory analyzed photometric observations made by Dale Cruikshank and Robert Brown in the summer of

STREAKS OF CIRRUS-LIKE clouds drift 75 to 100 kilometers above the main cloud deck of Neptune.

WINDS SURE BLOW FAST, WAY OUT THERE! Neptune turned out to have the fastest winds in the solar system.

1980; these gave "rotation periods" of 17.73, 18.29, and 18.56 hours. Since the planet's actual rotation wouldn't vary like that, the results pointed to wind-driven bright cloud features. Later Heidi Hammel of the Jet Propulsion Laboratory studied ground-based methane band images of Neptune and found additional evidence of atmospheric motions.

Nevertheless, the length of Neptune's day — the basic frame of reference — remained unknown even as the spacecraft was nearing the actual encounter. The problem was and is more than academic because models of the interior structure require a rotation rate of about 15 to 16 hours. The difference between this period and the

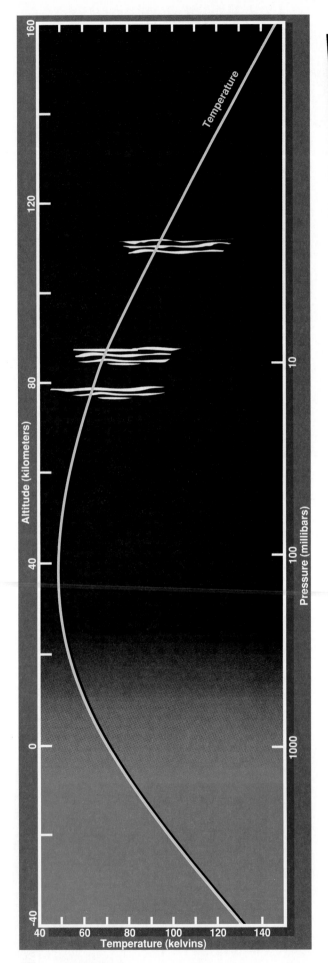

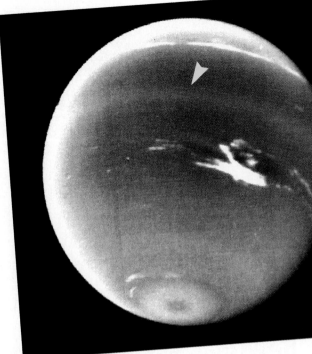

STREAKY AEROSOL CLOUDS FORM HIGH in Neptune's atmosphere (left) when ultraviolet sunlight breaks up ammonia and methane. Seen in the light of methane, a faint band marks Neptune's equator (arrow, above).

18-hour periods derived by Mike Belton from photometry pointed to very fast winds indeed. In the end, Voyager didn't pin down the length of Neptune's day until just before closest approach. From early results on the in-bound leg, many scientists predicted that the radio-emission period would be 18.3 hours, close to the rotation period for the Dark Spot.

Once the radio emissions were fully analyzed, however, scientists discovered the Dark Spot rotates more slowly than the deep interior by more than two hours every rotation. Thus it is actually drifting westward at a high rate. Translated into velocity, the measurement shows the Dark Spot lies where the wind blows westward at more than 2,000 kilometers per hour. (The diagram on page 41 shows the wind velocities at several latitudes, drawn from work by Lawrence Sromovsky and me.)

How can the Dark Spot — a fluid structure larger than Earth — do this? This was a question that gave quite a few sleepless nights to Verner Suomi, a scientist at the University of Wisconsin and a member of the imaging science team. In a recent paper published in the journal *Science* he and his colleagues suggest that the Dark Spot (and the rest of the atmosphere at cloud level in low latitudes) lags behind the internal rotation because their motions originate in the deep circulation within Neptune.

How does deep circulation link to a westward drift at the surface? Angular momentum is the key. If any material, such as a bubble of hot gas, rose up through Neptune's atmosphere, it would retain its original angular momentum. Brought to the surface, which lies farther from the planet's spin axis, the bubble would rotate more slowly, much as whirling ice

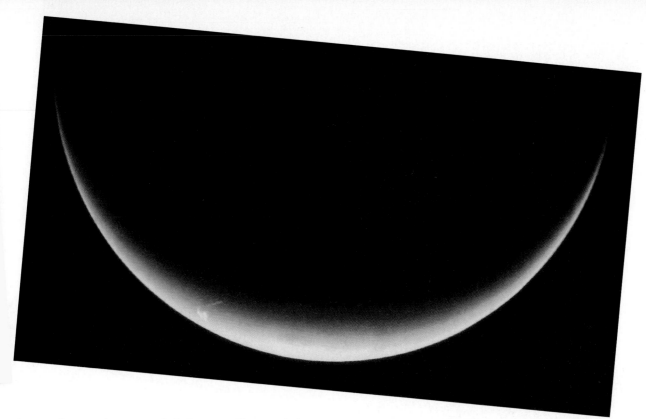

EVENING HAZES ARE BLUE (left side), morning ones are red. During the 9-hour Neptunian night, some unknown process alters the atmospheric aerosols or methane.

skaters slow as they extend their arms. The result is that the bubble would lag behind the surface rotation and appear to drift westward.

But does Neptune have rising parcels of gas? The answer clearly is yes. Measurements show the planet loses more heat than it receives from the Sun and so its surface is colder than its interior. In cooling, Neptune's atmosphere rises to the top and because Neptune is largely gaseous, it is quietly "boiling" like a giant saucepan. As upwelling parcels of Neptune's atmosphere rise from the deep interior, they lag in rotation speed. From the point of view of the visible surface, they move rapidly westward with reference to the deep interior rotation rate.

Neptune's wind profile is similar to that of Uranus: in the equatorial latitudes, winds blow against the sense of rotation and *with* it in the higher latitudes. The main difference is that the crossover point lies at lower latitudes on Uranus. According to models developed in our work, this implies that the zone of convection — the region of rising gas — extends to greater depths on Neptune. This agrees with the observation that Uranus loses less heat than Neptune, although they are about the same size.

Toward the Unknown Regions

Neptune's windy weather promises to help planetary meteorologists look inside the planet and puzzle out its heat "engine." On Earth, the equator is cooled and the poles warmed by the transport of heat within the atmosphere-ocean system and by water changing its phase from solid to liquid to vapor and back again. On the gaseous planets, however, the heat exchange occurs entirely within their deep atmosphere and with little change of phase.

We have learned that even when the heat absorbed from the Sun is significant, the tilt of a planet's axis is not particularly important. For example, both Venus and Jupiter have negligible tilts, so their poles receive little sunlight — yet their poles have almost

the same temperature as their equator. Uranus provides an extreme example: despite having one of its hemispheres in darkness for half its 84-year orbit, the temperatures at the dark pole and the sunlit one are nearly the same. Neptune, with a moderate axial tilt, is the same. Yet while its temperature differences are small in an absolute sense, they are nevertheless relatively significant because Neptune's temperatures are very low overall.

Surprisingly, Voyager's IRIS measurements found no trace of the Dark Spot. This might indicate the Dark Spot is covered with a haze that blocks infrared emission. It could also indicate that the Spot extends deep within Neptune to a place where the rotation period is shorter. Thus, the visible Dark Spot and its infrared counterpart may have different longitudes.

We have by no means finished analyzing the Voyager data, and much more remains to be learned from future work. Over its 12-year journey through the solar system, Voyager has made crucial contributions to our understanding of the weather on the gas giants. This era of continual discovery will remain unmatched in human history for a long time. We are so close to the events that it is difficult to place a value on this knowledge, but whatever scientists of the future may come to think about the project, I feel very fortunate to have been part of it. □

Sanjay S. Limaye is a planetary scientist at the University of Wisconsin. He has worked on the Mariner 10 mission and on Voyager. Recently, he has been engaged in a project to measure the environmental impact of the Kuwaiti oil-field fires.

Seeing a Star's Surface

Even the world's largest telescopes can't resolve details on the surface of a star other than the Sun. Instead, astronomers use a star's rotation to unveil its features.

by David Bruning

Every astronomer at one time or another has looked at a star through a telescope, trying to magnify the image more and more to glimpse the star's disk. And everyone becomes disheartened because no matter how hard they try, they can't see the star's disk. Even the closest stars, like Alpha Centauri, appear no more than one-hundredth of an arcsecond across, resisting the efforts of even giant telescopes to resolve the disk.

Of course, you don't need to see spots on the surface to deduce their presence. For several decades astronomers have followed variations in the brightnesses of some stars. Setting aside those stars that vary in brightness due to pulsations or eclipses caused by a companion star, the variations imply that dark spots and bright patches called plages, both caused by magnetic fields, pop in and out of view as the star rotates. But knowing that a star is magnetically active is vastly different from knowing the details of the star's activity.

Because the youngest stars are in general also the most active stars, learning about stellar activity will help astronomers better understand how stars age. Closer to home, solar activity is important because it's related to the heating of Earth's atmosphere and long-term changes in Earth's climate. To predict how our climate changes, scientists need a good model of how the Sun generates its magnetic fields. But so far these models have proven inadequate. Mapping the surfaces of stars with their dark spots is an essential step in building better theories about the generation of stellar — and solar — magnetic fields.

Imaging an Unseen Surface

To get stars to reveal what lies on their surfaces, stellar astronomers have resorted to trickery. They cajole the stellar secrets by studying variations in a star's spectrum and brightness, using the star's own rotation to betray the presence of dark spots, bright plages, magnetic fields, and regions of peculiar chemical composition. Doppler imaging, as researchers call it, uses the shifting in spectral-line wavelength caused by the star's rotation.

To see how Doppler imaging works, consider the surface or photosphere of our star. Rotation causes one limb, or edge, of the Sun to move toward us and the other to move away. Spectral lines emitted at the limb moving toward us shift toward the blue end of the spectrum, while lines emitted at the limb moving away shift toward the red. Between the two limbs, lines shift less to the blue and red, with the lines at the center of the Sun's disk not shifting at all. If you know how much a line has shifted and whether it shifted to the red or the blue, you can tell which region on the Sun emitted the spectral line.

If we saw the Sun the same way we do other stars, we would no longer see individual spectral lines emitted by the different regions. Instead we'd see all the redshifted and blueshifted lines blended into one broad spectral line. The faster the star rotates, the broader the line would be. Although you might think a disastrous muddle would result, the broad line still can reveal the location of a dark spot on the star's surface.

A starspot, being cooler than the surrounding gas in the star's photosphere, emits a slightly different spectral line. This difference appears in the broadened line as a "bump." As the spot rotates onto the visible hemisphere of the star, the bump first appears in the bluest part of the broad spectral line. As the spot rotates across the

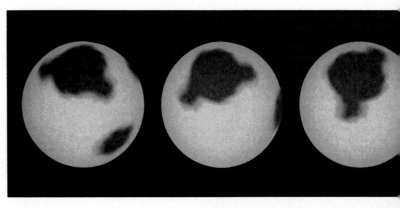

VIEWING DARK SPOTS on the surface of a star, such as solarlike LQ Hydrae, is no longer a dream of astronomers. A new technique called Doppler imaging permits them to map spots (black) and bright patches called plages (yellow).

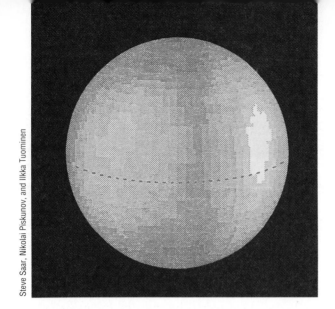

MAPS OF MAGNETIC FIELDS REVEAL that the strongest fields (green) on LQ Hydrae don't lie in the dark spot regions but in bright plage regions.

face of the star, the bump becomes less blueshifted as the spot approaches the center of the stellar disk. Then the bump starts to redshift as the spot moves toward the other limb and eventually disappears as the spot passes to the invisible side of the star.

By watching a star's changing spectrum, astronomers can trace bumps as they move through a line from blue to red. They know what positions in the line correspond to what longitudes, and the brightness of the bump combined with stellar brightness measurements helps decipher the latitude of the spot. By watching the bump as it goes from blueshifted to redshifted, the researchers can reconstruct with certainty the latitude and longitude of the spot — and the positions of other spots if they are present. (In practice, astronomers use several spectral lines to deduce the locations of spots to reduce analysis errors.)

This procedure requires lots of telescope time to obtain the precise spectra needed to make Doppler maps of the star's surface. If the star rotates quickly, observers can obtain enough observations to cover the star's surface in a few nights. Stars that rotate slowly are

A STAR'S ROTATION PERMITS astronomers to translate bumps in the star's spectrum into spot locations. Many stars, like HR 1099 shown here, have large spots at the poles, unlike the Sun, where spots group near its equator.

more difficult because their narrow spectral lines make it hard to follow the bumps.

The mapping process also favors stars with large spots because small spots make only tiny, unobservable bumps in the spectral lines. Also, stars that produce many small spots scattered fairly evenly over the surface do not show changes in the spectral lines. So spots on these stars are effectively invisible with present methods.

The high-precision spectrometers needed to make these observations have been available only in the last decade, so the number of mapped stars is low. The conditions needed to map spots — fast rotation rate, variations in brightness, and large spots — also have conspired to keep the number of mapped stars low. The best examples so far are the RS Canum Venaticorum (RS CVn) stars.

When Are Spots Like Zebras?

RS CVn (pronounced Ar-Ess-Kan-Ven or Ar-Ess-See-Vee-En) stars are named after the prototype of this group of variable stars, RS Canum Venaticorum. They are binaries; the primary star is a F-, G-, or K-type giant or subgiant star and the companion star is a normal G- or K-type main sequence or dwarf star like the Sun. The RS in RS CVn points out that these stars vary in brightness — not because the binary stars eclipse each other but because the primary star has large spots.

The spots appear to form near the equator and then slowly migrate toward the pole. This goes counter to the Sun's spots, which form at high latitudes and then migrate to the equator. The backward behavior and large size of the spots on RS CVns interest astronomers trying to explain how and why spots form on stars.

Astronomers have observed a number of RS CVns so far: El Eridani, HR 1099, UX Arietis, HD 12545, and AR Lacertae. Typically the spots on these stars are about 1000 kelvins cooler than the surrounding photosphere, roughly the same temperature difference on our Sun between sunspots and the surrounding photosphere.

El Eridani, a 7th-magnitude G-type subgiant, has provided an important test case for mappers. Four different research groups with different mapping techniques have mapped the surface of this star. While not all the methods produce exactly the same results, all showed polar spots. This dispelled some suspicions that the mapping methods weren't locating real surface features. Any method can produce some erroneous features, but the features shown by all four methods must be real.

Of course, the groups' efforts don't solve the so-called zebra problem. Are we looking at normal photo-

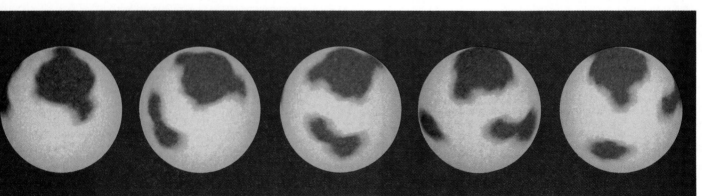

sphere over most of the star with a dark spot at the pole? Or are we seeing normal photosphere at the pole with a bright magnetic feature at the equator? White-on-black or black-on-white?

To solve the zebra problem, Larry Ramsey and Harold Nations of Pennsylvania State University searched HR 1099 (a 6th-magnitude K-type subgiant with a solarlike G2 dwarf companion) for other signs of starspots. They found that when HR 1099 showed its supposed spots, the spectral signature of titanium oxide was also present. Titanium oxide is a common ingredient in sunspots (and white paint) but it doesn't normally appear in RS CVn stars at all. So its presence confirmed that the stars indeed have dark spots.

Different Rotation

UX Arietis is another RS CVn star that has revealed an important clue as to how spots form on stars. Steve Vogt of Lick Observatory in California and Artie Hatzes, now at the University of Texas, tracked UX Arietis for five months and found several spots on its surface that changed during this interval. More importantly they could measure the rotation rates of the spots at several latitudes. Vogt and Hatzes found that the star rotates at different rates at different latitudes.

A solid ball rotates at the same rate for all latitudes. Every latitude of Earth, for example, rotates once every 24 hours. But the Sun rotates differentially, with regions at the equator rotating once every 25 days and regions near the poles rotating in about 35 days. UX Arietis rotates in about 6.4 days at the equator and slightly faster near the poles — its differential rotation goes opposite that of the Sun's.

Theorists believe that convection, the transport of energy by rising hot gases and falling cool gases, interferes with the normal rotation. The resulting differential rotation may help feed energy to the dynamo deep inside the star's interior that generates the star's magnetic fields. As important as differential rotation is, astronomers have seen it on only a few stars, including UX Arietis. Astronomers will need to observe differential rotation on many more stars before they will be able to decipher how rotation, convection, and magnetic fields interact.

Chemically Peculiar Stars

Spots on RS CVn stars aren't the only stellar features that astronomers have mapped with Doppler imaging methods. John Rice of Brandon University in Manitoba and William Whelau of the University of Western Ontario have found iron and silicon impurities across the surfaces of chemically peculiar A-type stars such as Epsilon Ursae Majoris and Theta Aurigae. Normally stars have their iron and silicon sprinkled uniformly through the atmosphere. But in peculiar A-type stars, some regions can contain up to 100 times more of these elements than other regions.

Astronomers believe the overabundance of these metallic elements in some regions results from the star's magnetic field. The star's gravity tries to pull these heavy elements toward the center while magnetic fields and light emitted within the star can force the elements to the surface. Scientists call this tug of war diffusion. Because the star's magnetic field, like that of a bar magnet, has two poles, diffusion forms rings of iron and silicon around the magnetic poles.

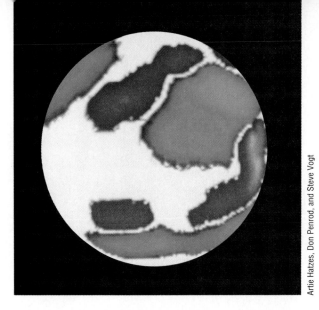

Artie Hatzes, Don Penrod, and Steve Vogt

MAGNETIC FIELDS CREATE PATCHES of iron and silicon (up to 100 times normal amounts), shown in red, on chemically peculiar A-type stars such as Gamma^2 Arietis.

Hatzes, along with Vogt and Don Penrod, has also looked at chemically peculiar stars, in particular 5th-magnitude Gamma^2 (γ^2) Arietis (see above). He found the magnetic field structure of Gamma^2 Arietis may be complex, which may yield important clues as to how magnetic fields aid diffusion and perhaps why these stars have strong magnetic fields.

And Dwarfs, Too

One dwarf star like the Sun has had Doppler images made of its surface, LQ Hydrae. Klaus Strassmeier of Universität Wien in Vienna, Austria, and his coworkers have mapped this 8th-magnitude K dwarf (see page 34). They find a number of spots that are a few hundred kelvins cooler than the star's photosphere, although the mapping method overestimates the temperature of the small spots of this star. Steve Saar of the Harvard-Smithsonian Center for Astrophysics and Jim Neff, now at Penn State, used titanium oxide spectral lines to determine that the spots are about 1200 kelvins cooler than the photosphere, which is similar to the difference in temperature between sunspots and the Sun's photosphere.

Saar, Nikolai Piskunov of the Astronomical Council in Russia, and Ilkka Tuominen of the University of Helsinki have extended Doppler imaging ideas to magnetic imaging to produce magnetic maps of LQ Hydrae (see opposite). These maps show that the strong magnetic fields lie near, but not exactly in, the low-temperature or spot regions. That suggests the magnetic fields relate more directly to the bright plages, as they do on the Sun, than the dark spots.

Astronomers have quickly moved from developing techniques for mapping stellar surfaces to confirming that what they see are in fact dark spots. They have mapped only about a dozen stars so far, mostly RS CVns and chemically peculiar stars, but are adding more stars all the time. Now researchers need to apply these methods to more solarlike dwarfs in their search to better understand stellar magnetic fields. Overall the tone of stellar astronomy has changed from "I wonder what the surface of a star looks like" to "I wonder what *that* star looks like." □

Some nearby stars such as Betelgeuse may explode in the next few thousand years, creating the most spectacular supernovae ever witnessed.

by Andrew M. Thorpe

Giving Birth to Supernovae

©Anglo-Australian Telescope Board

EXPLODING STARS such as supernova 1987A, which occurred in the Large Magellanic Cloud, tell astronomers about the synthesis of elements in the universe.

Exploding stars in other galaxies dramatically display stellar evolution in progress. The explosions, called supernovae, scatter gas and dust into the galaxy that later become the ingredients for new stars. Each successive generation of supernovae slowly builds the heavier and heavier elements, including those elements crucial to life as we know it.

Observing the bright outburst of light from nearby — and even distant — supernovae is relatively easy. But how does one detect a supernova before it explodes? Unfortunately astronomers can't predict when a star will explode, but they can predict which stars most likely will explode in the future. To identify these supernovae progenitors, researchers first looked at the types of stars that have exploded in the past.

Astronomers classify supernovae into four basic types: Ia, Ib, Ic, and II. A type Ia supernova originates in a close binary system when a star dumps mass upon its white dwarf companion, causing it to explode. A type Ib, Ic, or II supernova starts as a very massive star that collapses under its own weight once the nuclear fuel in the star's core runs out. The collapse of the star's core triggers the explosion of the star's outer layers into space. The spectra of types Ib, Ic, and II look similar except that Ib and Ic spectra lack strong hydrogen lines. Presumably, a companion star stripped away the progenitor star's hydrogen atmosphere before collapse.

So to identify supernovae progenitors, astronomers look for white dwarfs in close binary systems and for very massive stars. In both cases, such stars exist in our neighborhood of the Milky Way Galaxy. Some of these supernova progenitors are familiar to most backyard observers.

Betelgeuse and Red Supergiants

The most familiar example of a supernova progenitor is Betelgeuse, the bright red supergiant marking Orion's right shoulder. Located at a distance of about 650 light-years from Earth, this star is in the twilight of its life cycle. Different layers in the core fuse helium into carbon, turn helium and carbon into oxygen, and produce magnesium from carbon. Eventually, these processes will form silicon and iron. But then the fusion of elements will stop because, unlike all of the previous fusion steps, iron requires more energy to fuse than it releases. The stoppage of energy will cause Betelgeuse's core to collapse, triggering a supernova explosion. This explosion most likely will occur sometime during the next few thousand years because the supergiant phase of massive stars doesn't last very long.

Exactly when the explosion will occur depends in part upon the pulsations of Betelgeuse's surface. These pulsations currently give rise to changes in the star's brightness by up to a factor of 2, or almost one magnitude. The throbbing of the star's surface also mixes material from deep in the star with the hydrogen-rich material at the surface. The mixing can speed up or slow down the reactions in the interior. Also, the pulsations eject some of the surface material from the star into space. The loss of mass can cool the interior of the star, slowing the nuclear reactions inside.

Astronomers can't say yet whether Betelgeuse may eject enough matter to reduce or eliminate its chances of going supernova. At its present expenditure of over one trillion tons of material per second, much of the star's outer envelope will vanish into space. Without enough mass in the envelope to maintain its pressure on the core, Betelgeuse may suffer only a mild explosion, leaving behind a planetary nebula. The core would become a lowly white dwarf instead of the neutron star that a supernova would create. But if the mass loss rate slows and does not deplete the star's outer envelope by the end of Betelgeuse's supergiant phase, the core could still collapse, producing a spectacular supernova that would outshine the Full Moon in our skies.

Another factor makes scientists uncertain about Betelgeuse's future: a nearby companion star. The inner member of the system is so close to Betelgeuse that it actually orbits *within* the supergiant's atmosphere. During the remainder of Betelgeuse's lifetime, the gravity of the inner star may sweep up matter, further depleting the supergiant's hydrogen envelope. This companion could end up robbing Betelgeuse of the necessary mass needed for core collapse.

Using Betelgeuse as a standard, astronomers can derive criteria for other supernova progenitor candidates. Observations show that a progenitor loses mass and varies in brightness because of pulsations in the star's atmosphere. Pulsations dredge up elements from deep in the interior, so the outer envelope in the later stages contains heavy elements such as potassium, magnesium, and silicon. When the star creates these heavy elements, churns them up from the lower layers, and eventually ejects them into space, the supergiant is nearing its spectacular death.

These convulsions exist in other notable supergiants that may soon collapse into white dwarfs or neutron stars. Examples are Mira (Omicron Ceti), Ras Algethi (Alpha Herculis), Antares (Alpha Scorpii), and Mu Cephei (also called the Garnet Star). These stars are located within a few thousand light-years of Earth, undergo pulsations and mass loss, and exhibit brightness variations. Mira also shows signs of heavy elements in its outer envelope. These stars are all prime candidates for possible supernovae explosions in the next few thousand years.

Finding other red-supergiant progenitor candidates may be more difficult. These stars lie in the spiral arms of our Galaxy near the gas and dust clouds from which they formed. These same gas and dust clouds shield the stars from our view. But our search for progenitors has only started.

Blue Supergiants, Too

Though red supergiants are the most likely candidates to become type II supernovae, astronomers have recognized recently that some blue supergiants may go supernova. Supernova (SN) 1987A in the Large Magellanic Cloud resulted from the explosion of Sanduleak −69°202, a blue supergiant about 25 times the Sun's diameter. (Red supergiants are about 500 times the Sun's diameter.) Unlike supernovae from the explosions of red supergiants, SN 1987A did not appear as bright as expected.

Jim Baumgardt

MASSIVE RED SUPERGIANTS such as Betelgeuse (the reddish star at Orion's right shoulder) are prime candidates for future type II supernovae.

According to Kris Davidson of the University of Minnesota, "A red supergiant doesn't have to expand very much and it doesn't lose much energy before it becomes a supernova. But SN 1987A was a blue supergiant. By the time we saw it as a supernova, it had to expand a lot, wasting a lot of energy."

During the final stages of Sanduleak −69°202's life, the energy that normally would have made the explosion shine brightly went instead into the expansion of the gas, dampening the optical display. Astronomers are still debating the reasons, but the progenitor may have lacked the necessary elements, such as oxygen, required to create a stable red supergiant stage. Instead the star became a dense, blue supergiant. Its compactness tempered the brightness of the supernova's outburst.

Astronomers are watching other blue supergiants that may one day become supernovae. One candidate is Eta Carinae, the bright southern star located 9,000 light-years away in the vast gaseous cloud that bears its name. This star's brightness has varied dramatically over the last century and a half. In the 1830s and '40s, observers witnessed this supergiant rise from 4th magnitude to magnitude −1, making it the second-brightest star in the sky. Over the decades it faded slowly to 8th magnitude, but it is now back up to magnitude 5.8.

Astronomers now believe that the observed brightening was really a small explosion that engulfed the star in a nitrogen-rich cloud. Apparently, Eta Carinae is showing the symptoms of imminent disaster. Davidson says that Eta Carinae is now nearing its final stage as a supergiant and suspects that a supernova might occur within the next 10,000 years. With a mass that is 80 to 100 times that of the Sun, it may become the greatest supernova ever witnessed by human observers.

"It probably is about to become a Wolf-Rayet star," says Davidson. Wolf-Rayet stars are hot, massive stars that undergo intense mass loss. The large amount of nitrogen in their spectra shows that they have already become supergiants once — perhaps even red supergiants if massive enough. The large mass loss removes the star's outer layers, stopping the supergiant phase and turning it into a Wolf-Rayet star. "This kind of star can no longer be a red supergiant, so it becomes a Wolf-Rayet star before it goes supernova."

Eta Carinae's large mass, high rate of mass loss, and shell containing heavy elements meet the criteria for a supernova progenitor. Eta's mass promises a brilliant supernova display, easily outshining Venus at maximum light.

There are other Wolf-Rayet stars that are good candidates for type II supernovae progenitors. One is 11th-magnitude HD 56925 in the constellation Canis Major. The star is embedded in a bright nebula called NGC 2359. A second is HD 192163, which is a 7th-magnitude object in the Crescent Nebula (NGC 6888) in the constellation Cygnus.

Other Progenitors

Supernova progenitors are not always the most conspicuous stars in the Galaxy. In fact, type Ia progenitors may be beyond detectability. Type I systems normally are far less luminous than red and blue supergiants. Yet when they are at their critical phase, they can rival the light output of type IIs and even the brightness of their parent galaxies.

Type Ia supernovae occur in binary systems where one star loses mass — mostly hydrogen gas — to its white dwarf companion. Normally the accretion of hydrogen onto the white dwarf's surface sparks a small outburst when enough hydrogen accumulates to

THE BLUE SUPERGIANT Eta Carinae may explode in the next 10,000 years, creating the greatest supernova ever observed.

RED SUPERGIANT Antares (arrow) appears bright at the heart of Scorpius and someday may become a type II supernova.

start nuclear reactions. Astronomers observe these small outbursts as cataclysmic variables or novae. But in a type Ia supernova, the white dwarf receives too much mass from its companion. A typical white dwarf has 0.6 times the Sun's mass. The star can remain stable if its mass remains below the Chandrasekhar limit of 1.4 times the Sun's. Unfortunately, in a type Ia system the white dwarf accumulates more mass than this.

Instead of starting nuclear reactions at the star's surface, the gas causes the star to catastrophically collapse, forming a neutron star. Nuclear reactions start in the collapsing star's core. This sudden burst of energy causes the star to explode, annihilating itself.

An example of this process may be found in the constellation Scorpius. Astronomers have discovered what appears to be a type Ia progenitor approaching critical mass. U Scorpii is a variable star located about 8 degrees north of Antares near the Scorpius-Ophiuchus border. Astronomers first noticed this 17th-magnitude star in 1863 when an eruption brightened it. Since then it has brightened spontaneously up to 8th magnitude at least four more times. It is presently labeled as a recurrent nova that researchers believe is rapidly accreting matter from its companion.

Sumner Starrfield of Arizona State University believes that U Scorpii is reaching critical mass and should go supernova during the next 100,000 years. "Most of the material that is accreting from the secondary remains on the white dwarf. The mass is almost at the Chandrasekhar limit. It may take no time at all to drive it over."

Most cataclysmic variables and novae will not become type Ia supernovae. But Starrfield says the best candidates are recurrent novae. These stars may have accumulated enough mass such that they are near the Chandrasekhar limit. Finding these systems though is tough.

"Type Ia systems are often very faint," says Marshall McCall of York Observatory in Toronto. "It could well be that a system lies nearby and is just too hard to see."

Which Star is Next?

By astronomers' calculations, a supernova should occur in our Galaxy every 30 years on average. Many people believe this means a supernova near our solar system is long overdue. But the key phrase is "on average." Supernovae don't appear like clockwork. And there is a good chance that the next supernova will appear in a part of the Galaxy that astronomers can't observe.

Years ago some supernovae, such as Cassiopeia A, exploded without detection. Dust shielded the outburst from optical visibility 300 years ago. But modern radio and infrared observations will help prevent future supernovae from passing undetected.

Although researchers now have a better idea of which stars may end their life cycles in a supernova explosion, predicting whether a particular star will go supernova, much less when, remains difficult. The best candidate for a progenitor, Betelgeuse, has enough variables — mass loss, effects of companion stars, and internal mixing — that a clear prediction whether Betelgeuse will become a supernova still isn't possible.

So for the near future, astronomers need to be vigilant and survey the skies for those stars that have just become supernovae. Each star leads to a better understanding of which stars go bang in the night and what stellar characteristics indicate that a star is ready to go supernova. From this information, supernova predictions will someday be feasible. Until then, keep looking up, because no one knows which star is next. □

©Anglo-Australian Telescope Board

HOT, MASSIVE EVOLVED STARS that have lost their outer layers through mass loss may go supernova when their cores run out of nuclear fuels. The Wolf-Rayet star HD 56925 in NGC 2359 is one such star.

Andrew M. Thorpe wrote "Should You Keep an Astronomical Notebook?" for the January 1988 issue of ASTRONOMY.

BUBBLES in the Sky

Interaction between stars and the interstellar medium paints our Galaxy with bubbles and bow shocks. What creates these works of art?

by Dave Van Buren

Spiral galaxies are pretty smooth operators, at least viewed from afar. They appear to have a nice, orderly disk of stars, gas, and dust slowly pinwheeling around a bright center. At first glance the interstellar medium, the matter that fills the space between the stars, is pretty mundane — a thin, smooth soup of particles without apparent function in the grand scheme of the galaxies.

But a closer look at the interstellar medium reveals a more tangled reality. In the Milky Way Galaxy, for example, astronomers see extraordinary activity in the interstellar medium. Fierce stellar winds, hot temperatures, and stars traveling at high velocities create bow shocks and bubbles, or shells. Apart from being works of art on a cosmic scale, these unusual structures hold a great deal of information about how stars behave. And because stellar winds acting over long periods of time create many of these bubbles, the bubbles shed light on how stars lose mass over the course of millennia.

Our Galaxy is peppered with bubbles, which exist virtually anywhere intensely hot stars exist. Extremely hot stars with powerful stellar winds interact the most with the interstellar medium because of the winds and the effects of intense radiation. So scientists focus their research on two types of hot, energetic stars: O stars and Wolf-Rayet stars. O stars are normal stars with masses more than 15 times that of the Sun. In fusing hydrogen into helium at a terrific rate, these stars burn very hot, at temperatures above 30,000 K. They create winds that carry away up to 100 trillion tons of material per second at speeds of up to 3,500 kilometers per second. The force of these winds is so strong that it pushes the surrounding medium — gas and dust — out to distances of light-years away from the star. Some stars, such as Wolf-Rayet stars, have even more powerful stellar winds. These winds push tenuous material into shells, contributing even more matter into bubbles than the O stars do.

Just how does the Galaxy lend itself to bubble making? The scenario for creating a bubble is simple.

Suppose you have a newly formed O star sitting in a region of the Galaxy that contains dense interstellar gas and dust. At first, the star's wind rushes outward and sweeps the surrounding medium into a thin shell. Very quickly, after about a thousand years, enough

AN INTENSELY HOT WOLF-RAYET STAR created the bubble known as NGC 6888 in Cygnus (below). An X-ray image of NGC 6888 (bottom) made with the Einstein Observatory shows in false colors the highest energy regions of gas.

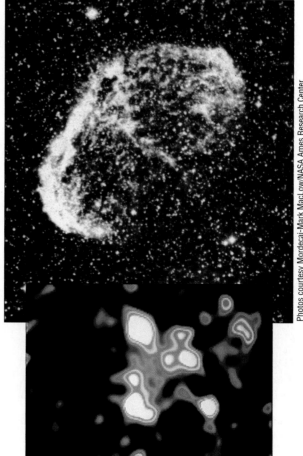

Photos courtesy Mordecai-Mark MacLow/NASA Ames Research Center

A GIANT COSMIC BUBBLE, the Rosette Nebula in Monoceros (opposite page) was blown by stellar winds and radiation from hot O stars in a central open cluster.

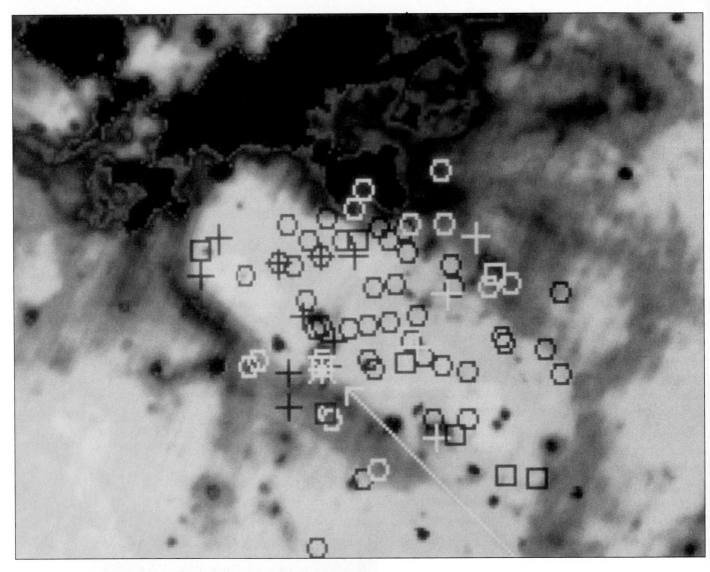

A HOT O STAR formed the shell known as the Bubble Nebula (NGC 7635) in Cassiopeia. It is bright enough and close enough to view with backyard telescopes.

THE CYGNUS SUPERBUBBLE, a shell spanning 400 light-years, is visible as the yellow oval in the center of this false-color infrared image. Crosses and circles show positions of O and B stars; squares show Wolf-Rayet stars. The arrow marks the open cluster Berkeley 87.

interstellar gas gets swept up that it begins to drag on the outer part of the expanding shell. At this point high-velocity wind from the star overtakes the shell and impacts it, creating a shock wave. As more and more gas is swept up, the shell's expansion slows and the material in the shell merges with the surrounding gas. By this time the bubble could be as big as 300 light-years across.

The largest bubbles are young ones surrounding Wolf-Rayet stars, extremely hot stars (50,000 K) that may be O stars stripped out of their outermost atmosphere. The Crescent Nebula (NGC 6888) in Cygnus, which was formed by a Wolf-Rayet star, is one of the brightest bubbles in the sky. Studies of the bubble show that most of the material in NGC 6888 was blown off the central Wolf-Rayet star during an earlier phase of the star's evolution.

Less exotic stars also produce bubbles. Run-of-the-mill O stars create bright, large bubbles, especially when they lie in thick parts of the interstellar medium. The Bubble Nebula (NGC 7635) in Cassiopeia was

PUNCHING THROUGH THE INTERSTELLAR MEDIUM at extremely high speed, the star Alpha Camelopardalis has created a bow shock nebula.

BUBBLES IN OTHER GALAXIES can be many times the size of those we know about in our Galaxy. This superbubble in the Large Magellanic Cloud surrounds a hot cluster of stars.

created by an O star. Because this star lies in a dense environment, the outside pressure against the bubble is great, keeping the bubble small. Because the bubble's expansion is small, the material pouring out into the bubble from the star reaches a greater density than it would if it expanded farther. Consequently, the bubble is relatively easy to see.

The Bubble Nebula is a case of an O star lying in relative isolation. But most O stars don't exist in isolation. Usually they form in groups. The Rosette Nebula (NGC 2237-9) in Monoceros is the bubble blown by a small cluster of O stars in the cloud from which they were born. The intense radiation from the stars is enough to excite material well past the bubble.

Don't imagine stars such as those in the Rosette cluster as sitting still in space. That's an over-simplification. On the contrary, most hot stars move rapidly through the interstellar medium. Some punch through the surrounding gas and dust at speeds of 100 km per second. These runaway stars, ejected from star clusters, do not create spherical bubbles but shells that resemble comets or the flow around a spacecraft reentering Earth's atmosphere. Such supersonic flows around a star are called bow shocks because they resemble the action of water as it flows past the bow of a ship.

Bow shocks, the ionized regions of matter surrounding high-velocity stars, are even more common than bubbles. The closest O star to Earth, Zeta Ophiuchi, has a prominent one, discovered by Ted Gull of NASA's Goddard Space Flight Center. Shortly after the Infrared Astronomical Satellite (IRAS) gathered data on many O stars, I studied a group of them and found a bow shock around Alpha Camelopardalis, another

runaway star. Because the star's wind is rushing outward in all directions, its ability to push material away decreases with distance. The interstellar wind of particles experienced by the star as it moves through space is steady. Because the star's wind decreases in strength with distance, a point exists around each of these stars where the stellar and interstellar winds are equally strong. Astronomers call this the standoff distance, where the bow shock occurs. The bow shock overtakes an unsuspecting parcel of interstellar gas, compresses and heats it, and makes it flow around the star. All the while dust in the gas absorbs starlight, heats up, and radiates energy in the infrared, becoming visible.

The Rosette Nebula, the Bubble Nebula, NGC 6888 — all of these objects are inhabitants of our Galaxy. Bubbles and shells, however, aren't limited to our Galaxy. Astronomers have observed superbubbles measuring hundreds of light-years across in nearby galaxies, including the Large Magellanic Cloud, the Andromeda Galaxy, M33, M101, NGC 55, and NGC 4449.

Studying how hot stars and star clusters interact with the interstellar medium is a new field. But as astronomers compile a record of more and more of these objects, analyzing them at different stages of evolution, they will discover a fuller picture of the story of how stars live and die. □

Dave Van Buren is an astronomer at the Infrared Processing and Analysis Center at the California Institute of Technology.

WHAT PUTS THE SPIRAL IN
SPIRAL GALAXIES?

After more than two decades of looking, astronomers finally have found traces of the waves that crash through the disks of spiral galaxies, creating the galaxies' characteristic pinwheel shapes.

by Debra Meloy Elmegreen
and Bruce Elmegreen

S ay the word "galaxy" and everybody — including astronomers — imagines a spiral galaxy. Of course, other types of galaxies exist, such as ellipticals and irregulars. But the arms that arc gracefully through the disks of spirals make these galaxies the most picturesque of all. The arms also make spirals the most enigmatic.

Spiral galaxies are bright, glittering deceivers. Like con artists working a fairground shell game, spiral galaxies divert your attention with the beauty of their swirling arms, while throughout their broad, thin disks, the real action goes on. For over a century, astronomers have theorized about what creates the pinwheel arms of spirals. Now observers think they have finally found evidence of the mechanism that shapes these galaxies: long-lived density waves that bounce back and forth in the disk.

Density waves act in some ways like waves on a lake. They sweep up matter into one location, leaving less matter nearby. But unlike lake waves driven by wind, the waves that shape a spiral arise from gravitational pulls within the galaxy. And unlike the patterns of lake waves that anyone can see, the subtle patterns of galactic waves long resisted astronomers' efforts to detect them.

When you glance at an image of a spiral galaxy like NGC 2997 (opposite), it's tempting to think the arms of the spiral directly trace the waves. As astronomers are now learning, the waves in a spiral overlap each other to form a more complex pattern that hides in the details of the arms' structure.

Finding evidence for density waves in spirals wasn't easy. First, astronomers had to rethink the way they classify galaxies by their shape. Then they had to process telescopic images of spirals in a new way to enhance their arms.

Spirals — the Inside Scoop

The Irish astronomer Lord Rosse first observed the structure of a spiral galaxy when in 1845 he viewed the Whirlpool Galaxy (M51) with his 72-inch reflector. Since then researchers have found that spiral galaxies account for about one third of the estimated 100 billion galaxies in the observable universe. Most of the rest are elliptical galaxies, which range in shape from nearly round to cigarlike.

Besides their numbers, spirals and ellipticals have other differences. Ellipticals have little or no gas and dust, which makes them incapable of forming stars. Spirals, on the other hand, have large amounts of gas and dust and produce a few new stars each year. Elliptical galaxies have more stars concentrated at their center than at the outer edges but otherwise have little structure. Spirals, however, show three distinct regions: a central bulge, a nearly spherical halo, and a flattened disk.

The central bulge typically spans a few thousand light-years. Because the bulge is basically round and contains old, reddish, low-mass stars, it resembles an elliptical galaxy. At the very center of a galaxy is the nucleus. Recent Hubble Space Telescope results lead astronomers to believe that the nuclei of some galaxies contain a mammoth black hole.

The halo of a spiral surrounds the disk and contains old stars similar to those in the bulge. Perhaps most of the stars in the halo lie in globular clusters, groups of a hundred thousand stars. The halo stars formed out of dense pockets of gas soon after galaxies began forming, about a billion years after the big bang. They retain the original spherical shape of the collapsing gas cloud that formed the galaxy.

Inside the stellar halo, the disk contains most of the galaxy's stars and presents the galaxy's characteristic spiral shape. The disk is flat because the spin of the original gas cloud prevented it from collapsing equally in all directions. Our Galaxy, which is typical of large spirals, has about 200 billion stars spread mostly through its disk, which is 100,000 light-years across and about 3,000 light-years thick.

As we look out from our position in the disk of the Galaxy, we see many stars along the plane of the disk but fewer stars perpendicular to it. When we look at the band of light in the night sky called the Milky Way, we are seeing the plane of our Galaxy.

There's more to the disk than just stars. The interstellar medium, or stuff between the stars in a galaxy, is mostly hydrogen gas. Gas accounts for about 5 to 20 percent of the visible

GRACEFUL SPIRAL ARMS DOMINATE OUR VIEW of spiral galaxies like NGC 2997 in constellation Antlia. Density waves sweep gas and stars in and out of the arms that arc through the spiral's disk.

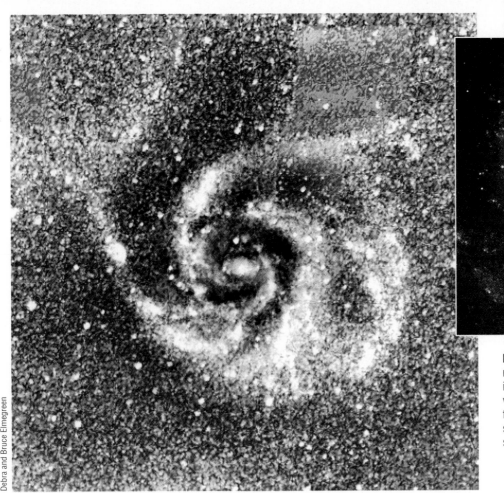

Debra and Bruce Elmegreen

INNER AND OUTER ARMS APPEAR with nearly equal contrast in this enhanced image (left) of M101 in Ursa Major. The view combines blue and infrared images to show light from both young and old stars. A visible-light photo (above) shows a normal view of the galaxy.

mass of a spiral galaxy, compared with less than 2 percent for ellipticals. Some of this gas diffuses through the galaxy, while other gas gathers into discrete clouds. These pockets of gas range from a few hundred solar masses in size to 10 million solar masses. The biggest clouds concentrate in or near the spiral arms; the smaller clouds scatter throughout the disk. Gas clouds are among the most important features of spirals because they are the birthplaces of stars.

From Gas Clouds to Stars

Collapsing gas clouds create stars with masses ranging roughly from one-tenth to forty times the mass of the Sun. Contrary to what you might expect, collapsing gas clouds favor making many small stars rather than one large star. Thus most stars in a galaxy are low-mass ones with temperatures under 5000 kelvins, which makes the stars appear orangish to reddish. Only a few stars contain much more mass than the Sun, but these powerhouses of energy have temperatures up to 40,000 kelvins and bluish-white colors.

Shining millions of times brighter than their low-mass cousins, high-

mass stars dominate the light in a galaxy way out of proportion to their numbers. But although these bright stars often lie in the arms, they don't directly trace the density waves. That's because the waves don't directly cause star formation. Instead the waves merely sweep stars and gas in and out of the arms.

All spirals have the same components of bulge, halo, and disk, and share the trait of star formation, but their shapes differ. Some spirals have bigger bulges than others. The shape of the arms also differs, some being wound loosely around the central bulge, others more tightly. To keep track of these different bulge and arm shapes, astronomers long ago devised a classification scheme.

Fat yet Fast

In the 1920s, Edwin Hubble first classified galaxies as S for spirals and E for ellipticals. He used two main criteria for subdividing spirals: the size of the bulge relative to the disk and the tightness with which the arms wrap around the bulge. Class Sa denotes galaxies with big bulges and tight arms, like the Sombrero galaxy (M104) in Virgo. Sc describes those

with small bulges and more open arms like the Pinwheel galaxy (M101) in Ursa Major (above). Sb forms the intermediate type for galaxies like M81 in Ursa Major (opposite).

In addition, Hubble noted, some spiral galaxies have bar-shaped configurations. He distinguished these, from the ordinary S galaxies by calling them SB, and subdivided them into a, b, and c, as before. SBa galaxies have bars that extend over one-third to one-half of the optical disk, whereas SBc galaxy have bars that reach only about one-fifth to one-third the disk size. Theoretical models indicate that the bar maintains its shape even after many rotations of the galaxy, much like a yardstick spun about its midpoint.

Since Hubble's work, other classification schemes have arisen, such as those developed by Sidney van den Bergh at the Dominion Astrophysical Observatory and Gerard de Vaucouleurs of the University of Texas. (De Vaucouleurs added subclass "d" for galaxies with extremely small nuclei, such as NGC 7793 on page 38.) But all these systems still describe only the general shape or brightness of the arms.

Debra and Bruce Elmegreen

COMPUTER ENHANCEMENT REVEALS the symmetric wave patterns that create the arms of spiral galaxies. The image of M81 in Ursa Major at right has been stretched on a computer to make the galaxy appear as if it were viewed face-on instead of in its normal tipped orientation (above).

Astronomers have found that the general arm shape results from the orbital speed of stars in the disk. Stars and gas have nearly circular orbits in the disk, but Sa galaxies have peak orbital velocities (300 kilometers per second) that are much higher than Sc galaxies (100 km/s) of the same brightness.

To better understand how the arms form, we have developed a new classification scheme that addresses the details of the spiral arms.

Looking at Spirals in a New Way

The heart of our system lies in assessing how symmetric and continuous the arms are. We call galaxies like the Whirlpool Galaxy, M51 (see next page), that have long and symmetric arms "grand design." Galaxies like NGC 7793 (next page), with short asymmetric pieces that merely give the illusion of a spiral pattern, are called "flocculent" because of their fleecy appearance. Intermediate examples, such as NGC 613 (next page), showing characteristics of both flocculent and grand design galaxies, we call "multiple arm" galaxies. These galaxies have two symmetrical arms in the inner part of the disk, but

their outer regions are highly branched and include many spiral arms and arm segments.

Overall, we find a galaxy's arm classification is independent of its Hubble type; each of the Sa, Sb, and Sc types can be flocculent, multiple arm, or grand design.

The details of arm structure depend partly on the presence of internal bars, as was first shown by John Kormendy of the University of Hawaii and Colin Norman of the Space Telescope Science Institute. Galaxies with grand designs tend to have either nearby companion galaxies or bars. When galaxies have both bars and companions, over 90 percent are grand design in type. (It appears that interactions between galaxies are often responsible for forming bars.) So isolated spiral galaxies tend to be flocculent whereas spiral galaxies in clusters, which have lots of neighbors and more possibilities for interactions, tend to have a grand design pattern.

Analyzing arm structure is straightforward for galaxies viewed from above the disk, but most galaxies appear tilted to our line of sight. We can simulate looking face-on at them

by stretching computer images until they appear round (since we know galactic disks are circular). We also enhance the images because the brightness of the disk falls off sharply from the center to the edge of the disk. Enhancing the images lets us examine both the bright inner regions and the faint outer regions with good contrast.

Remarkable structure that wasn't easy to see in the original images becomes obvious when enhanced. For example, in M81 (above) the two main arms start bright, become fainter partway out, and then become bright again in the outermost regions. This pattern, which we have noticed in several galaxies, forms the first solid evidence for the existence of spiral density waves.

Doing the Wave

The theory of density waves dates back to the ideas of Bertil Lindblad of Sweden in the 1940s. They were elaborated upon in the 1960s by C. C. Lin of the Massachusetts Institute of Technology and Frank Shu of the University of California. Theorists continue to refine these ideas, which basically describe how a galaxy's

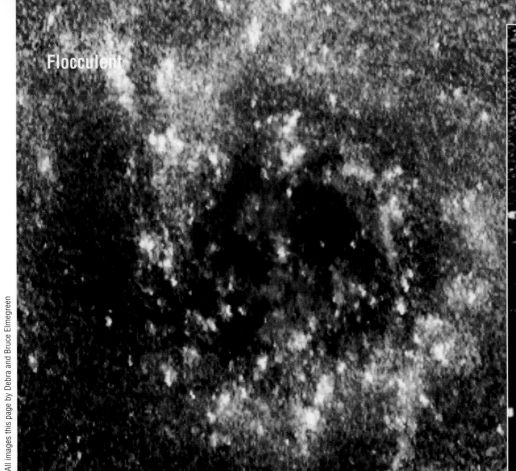

Flocculent

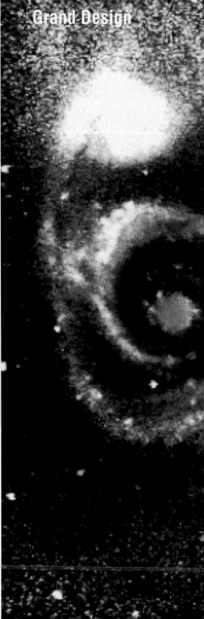

Grand Design

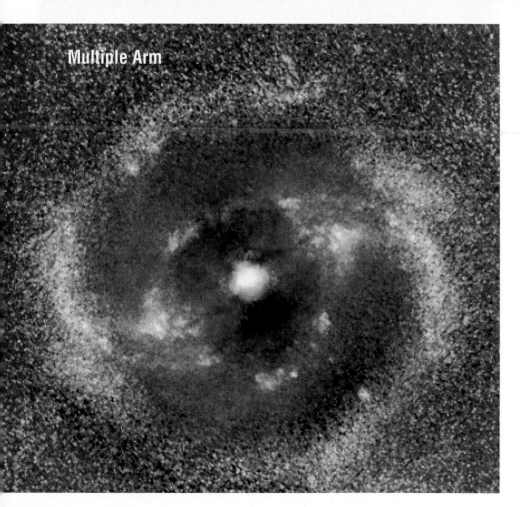

Multiple Arm

All images this page by Debra and Bruce Elmegreen

FLOCCULENT GALAXIES such as NGC 7793, an Sd galaxy in Sculptor, do not have large-scale density waves that organize the gas and stars in the disk into arms. Random regions of star formation give a broken arm appearance.

MULTIPLE ARM GALAXIES such as NGC 613, an SBb galaxy in Sculptor, have characteristics of both flocculent and grand design galaxies — symmetric arms and traces of random star formation throughout the disk.

GRAND DESIGN GALAXIES such as M51, the Sc-shaped Whirlpool galaxy in Canes Venatici, show symmetric arms. The arms glow with the light of young stars created by waves that compress the gas in the galaxy's disk.

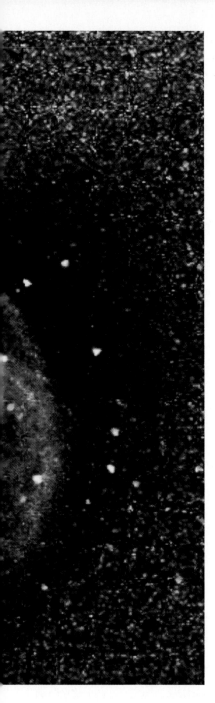

arms may result when star-forming regions, lying at different distances from the galactic center and orbiting at different speeds, temporarily line up to form an arc. This structure lasts only several tens of millions of years because the high-mass stars that illuminate the arm have short lives and rapidly fade from view. Computer simulations by Philip Seiden of the IBM Research Center and Humberto Gerola have shown how star formation can percolate through flocculent galaxies, creating the short, random arms characteristic of these galaxies.

Another difference between grand design and flocculent galaxies lies in the distribution of young and old stars through the galaxies. Flocculent galaxies appear different in red and blue light. Blue-light images show irregularly spaced arm segments created from newly formed stars, while red light shows a smooth disk of old, low-mass stars. The difference indicates that density waves haven't organized all the gas and stars in these galaxies.

In contrast, grand design galaxies have arms that stand out equally in red and blue light. This means that the density waves have organized all the gas and stars in the disk into a coherent pattern, and the exact nature of the waves determines the precise shape of the spiral arms.

Wobbling from Arm to Arm

Grand design galaxies also differ from multiple arm ones due to differences in the wave patterns within the galaxies. As each density-wave crest travels through the galaxy, it may encounter other waves. As the two pass through each other, they may add their energy or they may cancel, leading to a complex pattern of ripples through the disk. These wave patterns lead to the arm gaps and spurs and short spiral arm segments astronomers see in some galaxies.

Astronomers can trace these patterns by observing stars and gas. Stars and gas "wobble" about their nearly circular orbits around the galactic center because of the gravitational pulls of giant gas clouds and the spiral arms themselves. Astronomers call this periodic wobbling motion an "epicycle." If the time to complete one epicycle exactly matches the time it takes a star to go from one arm to the next, then a resonance develops. This is what happens when someone on a swing pumps her legs at just the right times to make the swing go higher. Resonance in a

galaxy means that stars soak up gravity wave energy and their epicycles become larger.

In galaxies with well-defined arms, the arms have an inner and outer limit beyond which they don't extend. (Sometimes the arms extend all the way into the galaxy's nucleus.) Inside and outside of these limits the galaxy's stars and gas completely absorb the energy of the spiral density waves and prevent the waves from going beyond to form arms.

Grand design galaxies have very prominent symmetric structures, with one side mirroring the other all the way from the inner limit to the outer. In these galaxies, each star completes its epicycle exactly once in the time between its encounters with each of the two spiral arms. This means it completes two epicycles during each orbit around the galactic center. Multiple-arm galaxies have prominent three-arm structures over part of their disks, indicating that three epicycles occur as each star completes its orbit around the center of the galaxy.

The rotation speeds of the gas and stars in the galaxy interacting with the angular speed of the wave pattern uniquely determine a galaxy's resonance pattern and thus the locations of its spiral arms. If researchers can find or guess the location of any one resonance (such as where the spiral arms end), they can then calculate the positions of the other resonances. So from prominent arm features, astronomers can discover the type of density wave pattern in any spiral galaxy.

So far astronomers have just glimpsed the barest of facts about density waves and how they shape spirals. They still need observations of many more galaxies in order to sort out the details of star formation and how these details relate to the presence of density waves. But it seems that spiral arm theories and observations finally are beginning to fit together, which has importance beyond explaining the shape of galaxies. These studies also form a key stepping stone for exploring how galaxies have evolved since their creation billions of years ago. ☐

gravity causes waves to ripple slowly through the matter in its disk. Wave crests temporarily concentrate gas and stars into regions that form the galaxy's arms.

As astronomers now see it, grand design and multiple arm galaxies have long, symmetric arms that are created by spirally shaped density waves. While individual stars may move in and out of the arms as they orbit the center of the galaxy, the wave pattern wheels around the disk at a constant rate. The relentless sweeping interaction of the wave with stars continuously piles up new gas and new stars to maintain the shape of the arm.

Flocculent galaxies seem to lack a global spiral density wave. Their short

Debra Meloy Elmegreen is an astronomer at Vassar College and Bruce Elmegreen is an astronomer at IBM's T. J. Watson Research Center in Yorkstown Heights, New York. They both specialize in the study of galaxies.

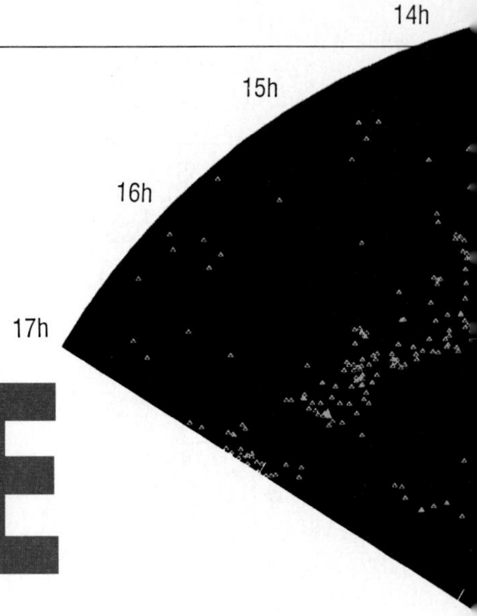

14h

15h

16h

17h

A New Map
of the
UNIVERSE

Astronomers have circumnavigated the universe in a project to map the structure of deep space.

by Alan Dyer

Imagine flying a billion light-years straight up out of the plane of the Milky Way. Our home Galaxy is now just a point in the distance surrounded by the specks of thousands of other galaxies. What does the universe look like from this perspective? Perhaps something like what we see in a new map produced by John Huchra, Margaret Geller, and Ron Marzke at the Harvard-Smithsonian Center for Astrophysics.

Their circular chart (opposite) plots the positions of more than 14,000 galaxies in a sweeping circle of space with us at dead center. The map encompasses a volume of the universe that extends 500 million light-years from Earth. By comparison the nearest large galaxy to ours, the Andromeda Galaxy, is 2 million light-years away. The most distant galaxies and quasars seen yet are 5 to 10 billion light-years away, roughly 10 to 20 times beyond the outer boundary of the new map.

Since 1985 the team has been mapping all the galaxies near us in space. They started with several 135°-wide by 6°-tall strips across the northern spring sky, surveys that produced the "wedge" style of maps such as the one shown above. They have now extended the survey completely around the sky. The latest map takes in all nearby galaxies within a swath that is 360° wide by

36° tall and extends from 8.5° north declination up to 44.5° north. If this were a map of Earth's surface, it would be like charting the surface all around the globe from a latitude of Panama north to Minneapolis.

The most striking feature of the all-sky plot is that it looks as if galaxies are lined up along spokes aimed directly at Earth. This "fingers of God" effect isn't real, however. How far a galaxy is away from the map's center is technically a measure of a galaxy's redshift, not distance. When a galaxy is moving away from us, its light waves become shifted to the red end of the spectrum. On the grand galactic scale of things, all galaxies are moving away from each other because of the expansion of space from the force of the big bang. The farther a galaxy is away from us, the faster it appears to be moving and the greater its redshift.

Therefore, redshift is more or less equivalent to distance. However, in the case of a tight cluster of galaxies, the member galaxies will be about the same distance from us but may have slightly different redshifts. This is because the galaxies are moving locally around the center of the cluster, motions that add or subtract from their redshift caused by the expansion of the universe. The resulting smearing of redshifts shows up as the radial spokes.

However, other features are real.

In the top half of the plot, you can see the so-called "great wall" of galaxies extending from 9 hours right ascension to 16 hours across the northern spring sky. The bright streak just above our location is the Virgo Cluster, a rich, nearby galaxy cluster well known to amateur astronomers as the home of many Messier objects. The streak above it at roughly 13 hours is the Coma Cluster, also known as Abell 1656. The clumping of galaxies in the bottom half from 23 hours to over 3.5 hours is the Pisces-Perseus Chain of galaxy clusters that dominates the northern autumn sky. The most prominent streak in that area (at 3 hours 16 minutes) is the cluster Abell 426. The two empty regions toward 6 and 19 hours are zones where our view to distant galaxies is blocked by the spiral arms of our own Milky Way Galaxy. Other circular empty regions, however, are thought to be real and give the universe on the grandest scale a soap bubble appearance. Why galaxies should congregate around the perimeters of empty voids is one of the mysteries that has cosmologists hard at work.

Meanwhile, Huchra, Geller, and Marzke are continuing their survey. By 1995 they hope to have completed mapping all galaxies brighter than 15th magnitude that lie in the northern half of the sky. □

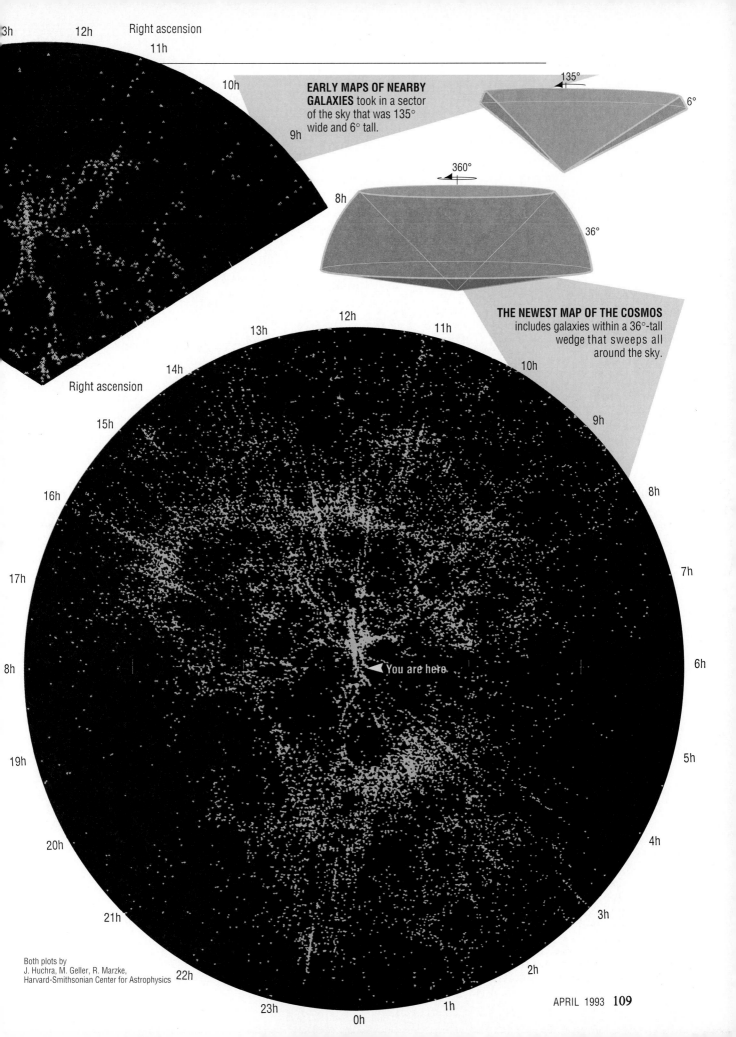

Right ascension

3h   12h   Right ascension
      11h

10h

**EARLY MAPS OF NEARBY
GALAXIES** took in a sector
of the sky that was 135°
wide and 6° tall.

9h

135°   6°

8h

360°

36°

THE NEWEST MAP OF THE COSMOS
includes galaxies within a 36°-tall
wedge that sweeps all
around the sky.

Right ascension

13h   12h   11h

14h   10h

15h   9h

16h   8h

17h   7h

8h   6h

You are here

19h   5h

20h   4h

21h   3h

Both plots by
J. Huchra, M. Geller, R. Marzke,
Harvard-Smithsonian Center for Astrophysics

22h   2h

23h   1h

0h

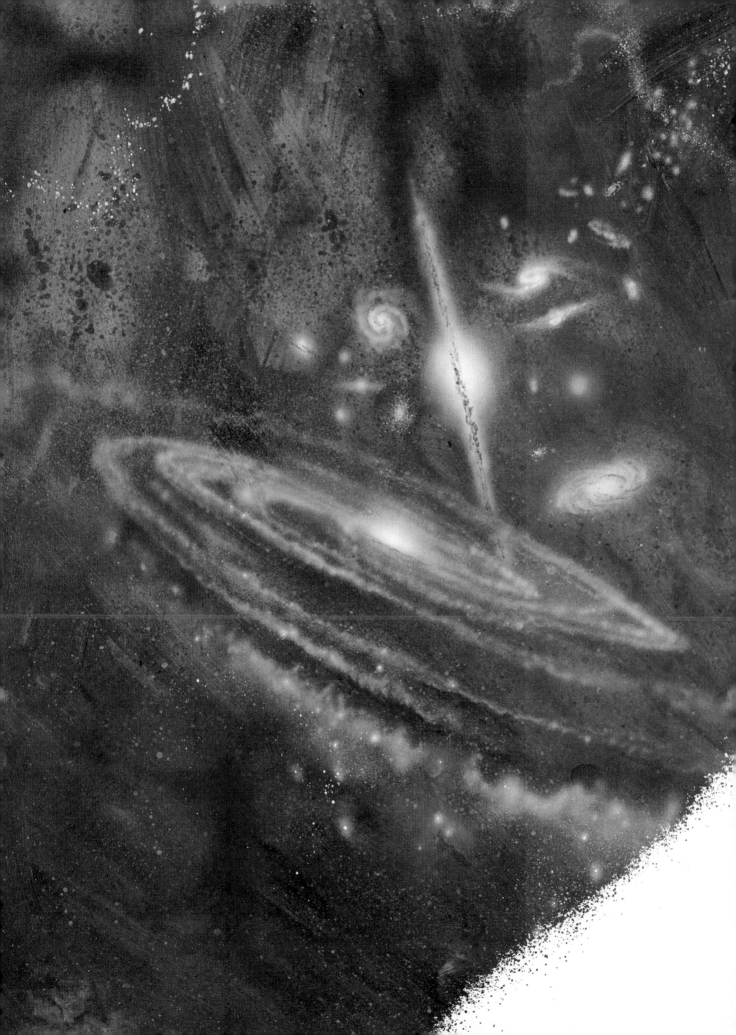

DARK MATTER PERMEATES the universe, from clusters of galaxies to the galaxies themselves. Even the destiny of the universe may be controlled by this unseen matter. Painting by Michael Carroll.

COSMOLOGY ————————

Shedding Light on Dark Matter

We can't see it, but we know it's all around us. It's called dark matter.

by Richard Monda

Fritz Zwicky uncovered a mystery in 1933, one astronomers are still trying to solve. Zwicky, a Caltech astronomer, found that the universe contains more matter than we see. This unseen matter deeply interests astronomers because its gravitational influence may control the destiny of the universe.

The first hint that dark matter exists came from observations of galaxy clusters. Zwicky measured the motions of galaxies in the Coma cluster and realized that the individual galaxies were moving too quickly for the cluster to stay together for a long period of time. The motion of each member of the cluster should have caused the cluster to fly apart long ago. Yet even backyard telescopes show that the cluster remains intact.

Zwicky concluded that the cluster must be ten times more massive than it appears in order for the cluster to remain gravitationally bound. This enormous mass discrepancy indicates that ninety percent of a cluster is invisible. Also, because matter in the universe clumps to form galaxies and galaxy clusters, the mass discrepancy implies that 90 percent of the *universe* is invisible.

At first, astronomers called the material "missing mass," but the term is misleading. Zwicky's observations show that the mass is there. "Dark matter" is a better term because it is the light emitted by the material that is missing.

There are cosmological implications to the existence of dark matter. The standard cosmological model predicts that the universe continues to expand and cool after the initial big bang. If, however, there is sufficient mass in the universe, the expansion will be slowed and possibly even reversed. Therefore, the amount of dark matter controls the expansion — and destiny — of the universe.

In spite of the obvious importance of dark matter, astronomers do not know the form of this matter. The list of discarded candidates gets longer each year, yet the form of dark matter remains murky. But we know where it is. It permeates the space between galaxies and the galaxies themselves.

Dark Matter in Galaxies

Just as the motions of galaxies in clusters indicate that matter lurks unseen in the cluster, the motions of stars give evidence for dark matter in galaxies. Vera Rubin, an astronomer at the Carnegie Institution of Washington, measures how quickly stars rotate about the center of galaxies. Plotting the stars' rotational speed versus their distance from the galactic center yields a rotation curve. Our Galaxy's rotation curve reveals that most of the galaxy — perhaps out to 300,000 light-years — rotates like a rigid wheel. Rubin's work conducted over the last two decades shows that other galaxies behave the same way.

Her results were surprising because astronomers expected stars to orbit a galaxy in the same way that planets orbit the Sun. Mercury, the innermost planet, speeds along at 30 miles per second, while the outermost planet, Pluto, plods at 3 miles per second. In contrast, stars in the Milky Way Galaxy move at roughly 150 miles per second, regardless of their distance from the center. The only possible explanation is that the visible part of the Galaxy is surrounded by large amounts of unseen matter.

Besides rotation curves, there is other evidence for dark matter. Astronomers find from the orbital motions of globular clusters that the Galaxy's mass is much larger than that of visible matter. The conclu-

sion is inescapable. Ninety percent of the mass of the Galaxy must be in the form of dark matter. The most likely location for this mass is in the halo surrounding the Galaxy.

The ratio of a galaxy's mass to its luminosity, or energy output, also supports the existence of dark matter. For stars like the Sun, the ratio is roughly one (one solar mass divided by one solar luminosity). Hot stars have ratios less than one because they emit lots of energy for their mass. Conversely, the ratio is greater than one for low-mass, faint red dwarf stars.

Visible matter in our Galaxy, which largely contains red dwarfs, has a mass-to-luminosity ratio of 2. But the ratio of total mass to luminosity is five times higher. The high total mass-to-luminosity ratio is found also in other galaxies and ranges from 10 to 30. Binary galaxy systems and clusters of galaxies have ratios as high as 300. Large

amounts of dark matter are needed to explain these ratios.

After 60 years of observation, in short, astronomers conclude 1) that galaxies and galaxy clusters must contain 10 times more matter than we can see and 2) that mass is most likely contained in the halo surrounding galaxies and in the spaces between galaxies. What isn't known, though, is the form of this invisible matter.

What Dark Matter Isn't

There are so many possible forms for dark matter that it is easier to state what it *isn't*.

We know it isn't atomic hydrogen gas because we would detect

A HIGH MASS-TO-LIGHT RATIO means most of the galaxy's mass is dark matter. M101 photo by Tony Hallas and Daphne Mount.

GALAXIES ROTATE at nearly constant speed rather than like the planets in our solar system.

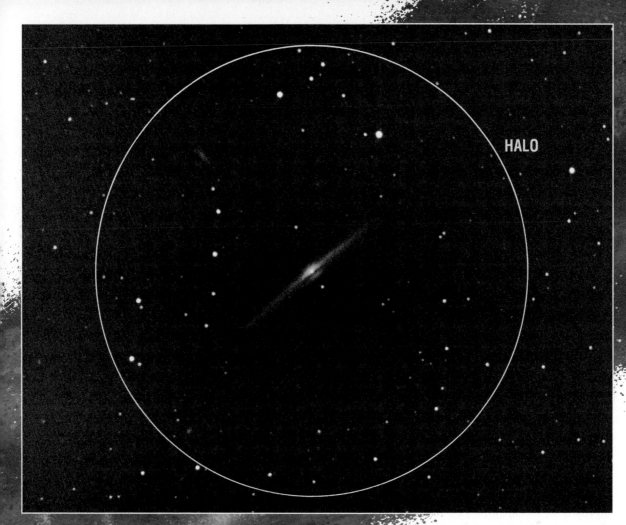

HALO

MASSIVE, FAINT HALOS SURROUND galaxies with dark matter. NGC 4565 photo by Jim Baumgardt.

radio emission from gas clouds. Likewise, optical radiation from ionized hydrogen gas clouds would betray their presence. Even molecular hydrogen would show absorption lines in the spectra of distant objects as light from these objects passes through the gas cloud.

We know it isn't dust, for large amounts of dust would obscure light from distant galaxies more than is observed. Dark matter can't be stellar mass black holes or neu-

tron stars either. Gas falling into these objects would emit more X rays than satellites have observed.

None of these radiation signals are detected by astronomers, so dark matter is not in the form of gas, dust, stars, or black holes. However, there are many other possible candidates such as brown dwarfs, black dwarfs, Jupiter-sized bodies, or exotic particles.

Brown and Black Dwarfs

Most stars visible in the Galaxy are faint, low-mass red dwarfs. But astronomers predict that even smaller objects, brown dwarfs, exist that are fainter and cooler than red dwarf stars. Brown dwarfs are at least 80 times more massive than Jupiter. They are more than just large planets, however. They form as stars do from large gas clouds

and produce energy for a short time through nuclear reactions.

To explain dark matter in the Galaxy by brown dwarfs alone, there would have to be at least 200 times the estimated number of red dwarfs. Far fewer probably exist, since the efforts of several groups of astronomers have produced only a handful of brown dwarf candidates (see "Do Brown Dwarfs Really Exist," April 1989). Some astronomers conclude that while brown dwarfs may contribute to dark matter, they do not exist in sufficient quantity to provide the large amount of observed dark matter.

Black dwarfs are another possibility. These objects are not failed stars like the brown dwarfs but stars that have run through their entire life cycle. A black dwarf is the ultimate endpoint for a low-mass star like the Sun. Such a star is black because it no longer emits light and, hence, is impossible to detect directly. But that isn't the biggest problem with black dwarfs: it takes too long to make one.

PLANETS MAY ROAM the galaxy, adding to its mass. Painting by Adolf Schaller.

FAILED STARS CALLED brown dwarfs each contain little mass but may be numerous. Painting by Mark Paternostro.

When massive stars "die," they quickly end up in their ultimate endpoint as a neutron star or black hole. But low-mass stars must first become a white dwarf. The white dwarf slowly cools as it gives off light and eventually becomes a black dwarf. Astronomers estimate it takes at least 10 billion years — the age of the Galaxy — for a white dwarf to become a black dwarf. Therefore, the Galaxy probably doesn't contain any black dwarfs, which turns our search for dark matter to yet another candidate.

"Jupiters" on the Loose

Planets should exist in the Galaxy in large numbers. Although most stars in the Galaxy are part of binary systems, many stars, like the Sun, are single. The formation process of these stars should have left large Jupiter-like planets orbiting them.

Despite attempts to detect these planets indirectly through motions of the star introduced by the planet's gravity, there are presently no confirmed detections of planets. (Direct sightings will be attempted with the Hubble Space Telescope when the telescope optics are corrected.) This suggests that either there are fewer planetary systems than first thought or the planets have low mass. In either case, planets alone are not the solution to the dark matter problem.

Exotic Particles

The absence of brown dwarfs, black dwarfs, and other planetary systems has driven astronomers to look for other dark matter candidates. One promising possibility is the high-energy particle "zoo." There are many contenders, such as photinos, gravitinos, and neutrinos.

The early universe was a sea of energy unlike today's matter-dominated universe. This energy-rich environment acted like an ultra-high-energy accelerator, creating nuclear particles we presently only envision. The high-energy phase of the universe's history ended less than a second after the big bang, but some of these exotic particles may still exist in the universe. Of particular interest is the class of particles called WIMPs. These Weakly Interacting Massive Particles are difficult to detect because they do not easily interact with other forms of matter. This low interaction makes them ideal candidates for dark matter.

Like other candidates for dark matter, most of these predicted particles have not been detected as yet. One exception is the neutrino. The Sun produces countless billions of neutrinos in its core as hydrogen is converted into helium. However, astronomers detect fewer neutrinos than predicted by models of the Sun's interior. One possible explanation is that the neutrino has mass — like protons and electrons but unlike the massless photons — and, therefore, interacts more strongly with other material inside the Sun than theory predicts. Currently, though, researchers believe that even if the neutrino has mass, it is too small to contribute significantly to dark matter.

Astronomers and physicists have joined in the search for other predicted particles. But in contrast to other candidates, such as brown dwarfs, not having detected these particles keeps interest high. Their large mass, potentially high abundance in the universe, and lack of interaction with visible matter (which keeps them in the dark) make these particles prime candidates for dark matter.

Cosmology and Dark Matter

Despite the many possibilities, astronomers still do not know the form of dark matter. But even without knowing its form, the cosmological implication of dark matter is clear. If there is sufficient material, the mutual gravitational attraction of objects in the universe will eventually halt the present expansion and cause the universe to collapse. On the other hand, if there is not enough material, gravity is too weak to slow the present expansion and the universe will expand forever. If the amount of mass is just right, gravity will bring the expansion to a halt but will be too weak to make the universe collapse. The expansion rate in this last case is called the critical speed.

The visible mass contained in the universe is 1,000 times less than needed to halt the expansion and even including present estimates of dark matter, the mass is still 5 to 10 times too small to cause the universe to collapse. To cause the collapse, the mass-to-luminosity ratio of galaxy clusters must be at least seven times greater than the largest ratio presently observed. Yet inflationary models of the universe predict that the expansion rate should be nearly the critical speed. Perhaps there is even more dark matter yet to be found in the universe.

How pervasive is dark matter? It is directly detected in galaxies and galaxy clusters. It might also exist in the space between clusters of galaxies and in the space between clusters of clusters. Astronomers have detected voids in the larger fabric of the universe where galaxies are not found. If dark matter exists in these regions, a major problem in astronomy may shift from how matter clumps in otherwise uniform space to what causes galaxies to form only in some regions of space.

As crucial as unseen matter is to our understanding of the size, shape, and mass of the galaxies and to the structure and destiny of the universe, we know only where it is located. There doesn't seem to be enough of it to stop the expansion of the universe and we know almost nothing about its form. Only one thing is really clear at this time. Astronomers are just beginning to shed light on dark matter. □

Richard Monda is the director of the Schenectady Museum Planetarium in New York.

EXOTIC PARTICLES LEFT OVER from the Big Bang may be the source of the dark matter.

ASTRONOMERS PRESENTLY BELIEVE there isn't enough mass in the universe, even with dark matter, to stop its expansion. Painting by Victor Costanzo.

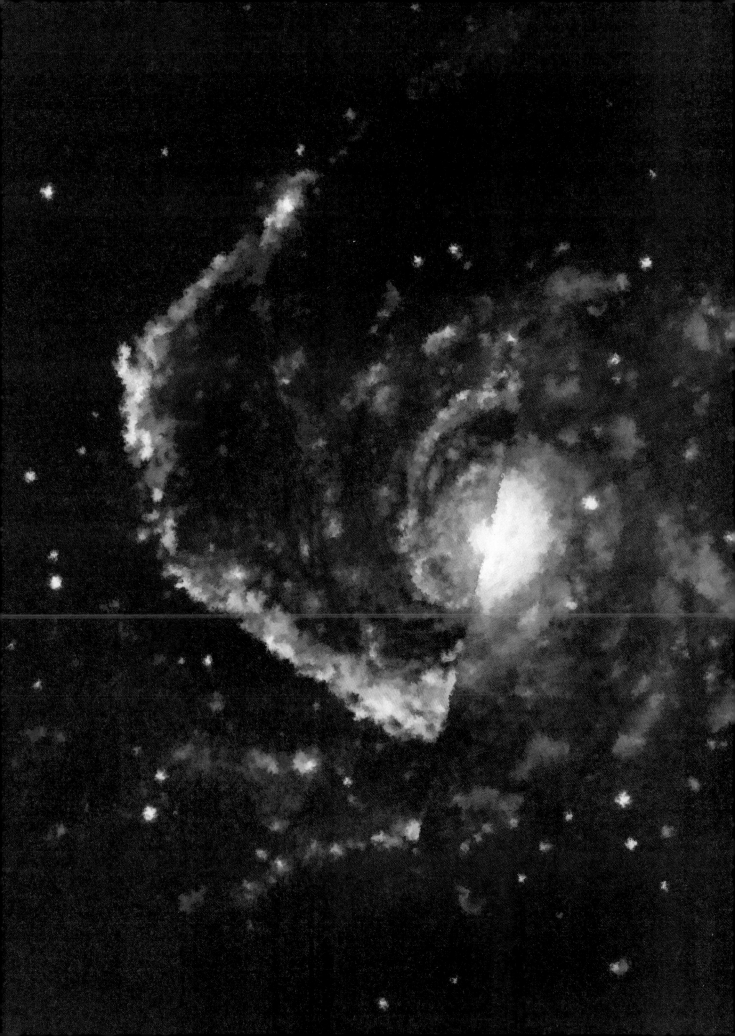

THE AGE PARADOX

Controversy surrounds a basic fact about the universe — its age. Stellar astronomers claim the most ancient stars appear to be several billion years older than the universe itself.

by Ray Jayawardhana

Cosmologists say the universe may be 8 to 15 billion years old. Stellar astronomers disagree. They say the oldest stars are much older, perhaps 16 to 19 billion years old. Because the oldest stars can't be older than the universe in which they lie, this age paradox presents a thorny problem for astronomers.

Resolving the age paradox is vital to understanding how the universe formed and evolved, but astronomers are having difficulty pinning it down.

Either the galactic distance scales used to find the age of the universe or the stellar evolution models used to find stellar ages must be wrong. Or perhaps both sets of data have flaws, as recent evidence suggests. Uncovering the flaws will provide astronomers with new information about how stars age, the distances to galaxies, and the expansion of the universe. Of the two ages, the age found from galactic motions, the cosmological age, is harder to determine.

ASTRONOMY: Thomas L. Hunt

GALAXIES PRESENT BOTH SIDES OF THE AGE PARADOX. The blue side of this galaxy reflects the young 8-billion-year age of the universe found by cosmologists from the motions of galaxies. The yellow side represents the 16- to 19-million-year age of the universe found by stellar astronomers for the most ancient stars in galaxies. Galaxies can't be young and old at the same time, so which age is correct?

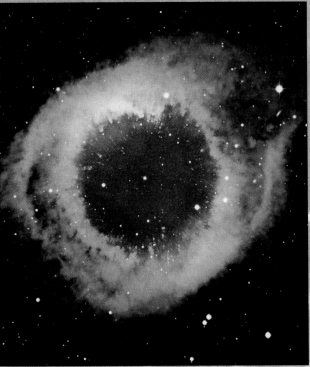

NEW STEPS IN THE DISTANCE SCALE come from bright, large planetary nebulae (similar to the Helix Nebula above) in distant galaxies, such as M83 (right). They confirm the distances found from other techniques.

Finding the Cosmological Age

Present measurements of the age of the universe had their beginnings in the 1920s when Edwin Hubble discovered that the universe is expanding. He observed that galaxies move away from each other and that more distant galaxies travel faster. But galaxies don't actually move through space. They are being pulled along with space as space expands. This expansion was caused by the big bang, the massive explosion that formed the universe.

Cosmologists determine the age of the universe by measuring apparent galactic motions. This method is akin to finding the time it takes to drive from one city to another by dividing the distance by the car's speed. Likewise astronomers find a galaxy's travel time since the big bang, which is equal to the age of the universe, by dividing how far the galaxy has traveled by its speed.

Errors in a galaxy's speed or distance lead to a wrong value for the age, so astronomers use instead the average expansion rate found from many galaxies. They find the expansion rate, called the Hubble constant (H_0, pronounced "H-naught"), by dividing the speeds of galaxies by their distances. The units of H_0 are unusual: kilometers per second per million light-years. For example if H_0 equals 30 kilometers per second per million light-years, a galaxy 100 million light-years away will recede from us at 3,000 km/s, 10 times faster than a galaxy 10 million light-years away traveling at 300 km/s.

One divided by the Hubble constant ($1/H_0$) gives an estimate of the age of the universe that astronomers call the Hubble age. This is the age that the universe would have if it contained no matter.

But stars and galaxies fill our universe. So its age must be less than the Hubble age because the gravitational pulls of the stars and galaxies slow the expansion over time. The slowing expansion means that galaxies traveled faster in the past than they do today. The faster-moving galaxies of eons ago could travel the same distance in less time than the slower-moving galaxies of today. Thus the Hubble age, which uses today's expansion rate in its calculation (the "naught" in H_0 signifies the present value of Hubble's constant), leads to an upper limit to the age. The actual age is less, probably around two-thirds of the Hubble age.

How can the oldest stars in our Galaxy be older than the universe?

Climbing the Distance Ladder

The Hubble age, and other estimates that derive from it, are only as accurate as the galactic speeds and distances that astronomers use to measure it. Finding a galaxy's speed is a fairly easy task. Astronomers use the Doppler shift: movement of an object away from us shifts the dark lines in its spectrum to the red, while motion toward us shifts the lines to the blue. The amount of shift in a galaxy's spectrum is proportional to its speed. Almost all galaxies show an apparent redshift because the universe is expanding.

On the other hand, measuring a galaxy's distance is a tricky business. Finding the distance to a galaxy involves using a series of steps that astronomers call the cosmic distance ladder, or scale. The first rung of the

ladder starts with the Earth-to-Sun distance and builds outward to the distances of other planets and nearby stars. The highest rungs rise out to the farthest galaxies and quasars.

Intermediate steps use "standard candles," objects having the same luminosity, to measure distances to distant stars and nearby galaxies. Knowing how bright the standard candle really is and comparing that with how bright the standard candle appears in a galaxy enables researchers to determine how far away it is. Previous steps in the distance scale help establish the distances found with subsequent ones, so astronomers leapfrog their way out to ever more distant galaxies.

Cepheids, a special breed of supergiant stars, are one of the standard candles used along the bottom rungs of the distance ladder. These stars vary regularly in brightness as they expand and contract in size. The speed at which they change brightness relates directly to their luminosity. So astronomers can deduce a Cepheid's luminosity by measuring how quickly it changes, and then find its distance by comparing its luminosity to how bright it appears. Because these stars never get extremely bright, they are useful for finding distances to relatively nearby galaxies, out to about 25 million light-years.

For more distant galaxies, researchers must use a brighter standard candle. Type Ia supernovae qualify as good standards because they all attain the same peak luminosity, which is several magnitudes brighter than Cepheids. When these stars collect enough material from a companion star, they explode violently and brighten rapidly, rivaling the brightness of the galaxies in which they lie. They can be seen out to distances of a billion light-years or more.

Unfortunately, stepping between the rungs represented by Cepheids and type Ia supernovae isn't easy — there's a hole in the ladder. Astronomers see Cepheids only in relatively nearby galaxies. But type Ia supernovae occur rarely, so observers hardly ever see these stellar outbursts in nearby galaxies. Instead they must look for supernovae among the larger number of distant

THE VIRGO CLUSTER PROVIDES a test of the methods used to measure galactic distances. Virtually all methods now yield the same distance to the cluster of galaxies and give an age of 8 to 12 billion years for the universe.

galaxies, creating a gap between the Cepheid and supernova distance scales. A further problem with supernovae is that the peak luminosity they reach is uncertain, making their distances imprecise.

The hole in the ladder has forced astronomers to use other methods to bridge the gap. The so-called Tully-Fisher method uses the fact that more massive galaxies are generally more luminous and rotate faster than less massive galaxies. Radio astronomers measure a galaxy's rotation rate from the width of the 21-centimeter spectral line emitted by the galaxy's hydrogen gas. The wider the line, the faster the galaxy is rotating and the more massive and luminous it is. Comparing this estimated luminosity with the observed brightness gives astronomers a measure of the galaxy's distance. The Tully-Fisher method yields values for the Hubble constant between 15 and 30 kilometers per second per million light-years, corresponding to ages from 14 to 7 billion years. But these results are too imprecise. Astronomers need more precise measurements of the Hubble constant in order to determine accurately the distances of faraway galaxies and the age of the universe.

Building New Rungs

Within the last five years, astronomers have used a new standard candle with great accuracy. They use planetary nebulae, large shells of glowing gas found around old stars. These dying stars kick off shells of gas because they are not massive enough to explode as supernovae. A team led by George Jacoby of Kitt Peak National Observatory found that the brightest planetary nebulae in galaxies make ideal standard candles

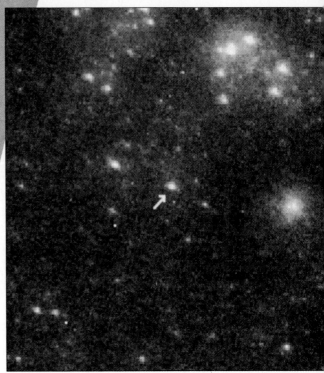

THE UNIVERSE MAY BE 15 BILLION YEARS OLD, as suggested by observations of Cepheid variable stars in the galaxy IC 4182. Enlarging the bottom of the box in the image above shows one of the Cepheids studied (arrow, above right).

because they all have nearly the same luminosity and are bright enough to be seen at great distances. These objects are especially useful because they usually lie far from the crowded central regions of galaxies and therefore can be observed easily. Also a unique signature in their spectrum helps to identify them.

Jacoby's team has established the distance to the nearby Virgo cluster of galaxies using planetary nebulae and found that their results agree well with those obtained from other methods. Following this confirmation, several research groups rushed to use the new cosmic yardstick. They found an age of 8 to 12 billion years for the universe.

Cosmologists strive to develop new rungs in the distance scale and to determine the Hubble constant more precisely. But despite their efforts to resolve the age paradox, cosmological ages hover around 10 billion years. The universe still appears much younger than the 16- to 19-billion-year age of the oldest stars.

Dating Ancient Stars

Stellar astronomers tell us that the oldest stars visible today lie in globular clusters and are around 16 to 19 billion years old. Globular clusters are giant collections of up to a million stars that orbit the center of our Milky Way Galaxy but reside outside the Galaxy's disk. All the stars in a globular cluster formed at the same time, so they have the same age. But massive stars in the cluster burn up their nuclear fuel faster than smaller stars, and thus become giant stars sooner. Astronomers observe which cluster stars are now evolving into giants and then use theoretical estimates of how long it takes

for these stars to become giants to find the cluster's age.

Looking at just one star in a cluster can give an erroneous age for the cluster. So astronomers plot the temperatures and luminosities of many stars and compare them with theoretical estimates of what the temperatures and luminosities should look like for different ages.

Don VandenBerg of the University of Victoria, Peter Stetson of the Dominion Astrophysical Observatory, and Michael Bolte of Lick Observatory have studied several globular clusters. They have estimated an age of 15 to 17 billion years for stars in the globular cluster M92, in Hercules. In fact M92 may be even older. An Italian group led by Oscar Straniero and Alessandro Chieffi of the Institute for Astrophysics in

Astronomers argue both ages can't be right. But which one is wrong?

Frascati thinks it may be around 19 billion years old.

These ages conflict drastically with estimates of the cosmological age of the universe. But stellar evolution theories may contain flaws, causing astronomers to believe globular-cluster stars are older than they really are. Yale University researchers including Pierre Demarque, Marc Pinsonneault, and Brian Chaboyer recently have studied a stellar process that previous models did not include. Diffusion within stars causes heavier elements such as lithium to settle closer to the star's center than the lighter ones such as hydrogen. This process changes how the star transports energy from its center to its surface and thus changes the temperature and luminosity of the star. So diffusion has a noticeable effect on the aging of stars and causes astronomers to misgauge the age of a cluster from stellar temperatures and luminosities.

Including the effects of diffusion in new evolutionary models, Demarque's group found that the oldest globular cluster stars are at least 16 billion years old. Unfortunately this leaves the age paradox intact. And a discovery made by another Yale astronomer makes things worse. Young-Wook Lee looked in the central bulge of our Galaxy at the oldest RR Lyrae stars, variable stars similar to Cepheids. He found that these stars are about 1.3 billion years older than the oldest globular cluster stars. His discovery pushes the age of the oldest stars to at least 17 billion years.

Although errors exist in the stellar ages, increases in age so far seem to offset any decreases in age. While further decreases may be possible as theorists make advances in understanding stellar evolution, the oldest stars appear no younger than 16 billion years. At present it seems that improved studies won't result in stellar and cosmological ages' meeting halfway. But some recent results from the Hubble camp point to a possible resolution of the age paradox from the cosmological side.

Resolving the Paradox

In June 1992 a team of astronomers from the Space Telescope Science Institute and the Carnegie Observatories announced a result that may fill the hole in the distance ladder. The team, which included Allan Sandage, used the Hubble Space Telescope to observe 27 Cepheids in a faint galaxy in Canes Venatici known as IC 4182. They picked this galaxy because astronomers had observed a type Ia supernova in this galaxy in 1937.

The earlier study resulted in an inaccurate distance to this galaxy because the supernova's peak luminosity was uncertain. But Sandage's team used the yardstick provided by the new Cepheid observations to calibrate the peak luminosity of the supernova and to determine a precise distance to the galaxy. Using this new distance, the team found for the Hubble constant a value of 14 kilometers per second per million light-years. This corresponds to an age of 15 billion years and comes close to resolving the age paradox.

While Sandage says that he is confident of the result, many others counsel caution. Michael Pierce of Kitt Peak National Observatory points out that IC 4182 contains a lot of dust, which could make the Cepheids look fainter and thus confound distance estimates. If Pierce is right, the Hubble constant derived for this galaxy would be larger and the age of the universe smaller than determined by Sandage.

And what of the Hubble constant values found for the Virgo cluster? The conflict between these values and that of Sandage's team means that the search for a cosmological age is far from over. Resolving the cosmolog-

National Optical Astronomy Observatories

THE OLDEST STARS APPEAR OLDER than the universe. Estimates of the age of the globular cluster M92 make it 16 to 19 billion years old. And some stars near the center of our Galaxy may be even older.

ical age problem may come from observing more galaxies. According to Jeremiah Ostriker of Princeton University, galactic motions independent of the expansion of the universe could confuse H_0 measurements. The way to avoid this problem, he says, is to use a large number of galaxies to calculate H_0.

Even if astronomers can agree on a cosmological age, the age paradox may remain — for a while. But most astronomers believe that as they use more reliable distance indicators and further improve stellar evolution theory, the cosmological and stellar ages will come into agreement. After all, we all know stars can't really be older than the universe itself. □

Ray Jayawardhana, an astronomy student at Yale University, has written for Science.

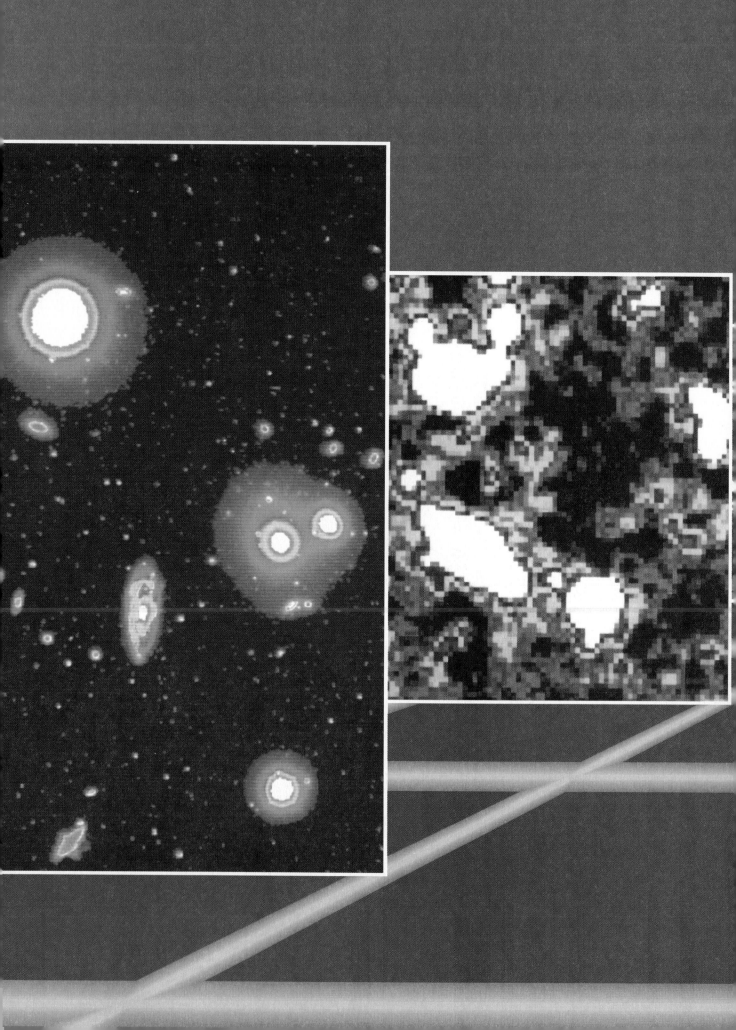

Beyond the Big Bang

New observations cast doubt on conventional theories of how the universe formed.

By Jeff Kanipe

BEYOND THE VISIBLE HORIZON of the universe, beyond the most distant galaxies and quasars we can see, lies the smooth and ancient background glow from the Big Bang. Astronomers want to know how the patterns of galaxies formed out of that smooth texture of energy and matter. Illustration by Steve Davis.

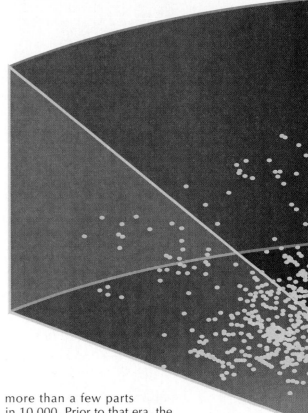

On November 18, 1989, the Cosmic Background Explorer (COBE) was launched into Earth orbit. Its mission: to study the dominant form of radiation in the universe, the cosmic microwave background that is believed to be the remnant heat from the Big Bang itself.

In just nine minutes of observations, COBE determined that the cosmic background radiates at a temperature of just a little over two and a half degrees above absolute zero. More importantly, it showed that the intensity of the background across the spectrum precisely matches that of a perfect emitter and absorber of radiation — an idealized object physicists call a "blackbody."

The COBE results were a stunning victory for the standard Big Bang model of the universe. Here was evidence that the primordial fireball was a uniform explosion of matter and energy. Out of a quantum seed the universe unfolded with the grace of a tropical blossom, balanced and smooth, but with all the structural attributes of a seaside fog. For astronomers, however, the COBE results added to a growing paradox in cosmology.

Paradox Found

The problem stems from COBE's confirmation that the background radiation is coming at us with equal intensity everywhere we look. This demonstrates that the early universe must have been smooth — matter and energy were once evenly distributed.

This is hardly the case today. You need only walk out into your backyard at night and explore the sky with a telescope to see that for yourself. The local universe is not smooth. Neither is the more distant universe we can see with giant telescopes. Lumps of matter exist in the visible universe — they're called galaxies and galaxy clusters.

Astronomers have mapped thousands of galaxies in three-dimensional space from both the Northern and Southern Hemispheres and have discovered that on large scales the universe is a vast, frothy tangle of galaxy clusters interspersed with voids of seemingly empty space.

As we peer farther into space and further back in time, we still find galaxies. One to two billion years after the Big Bang, astronomers see galaxies shining with the blue light of hot young stars. Even further back, shining like airport beacons from the very edge of the visible universe, astronomers still see lumps of matter — quasars.

The COBE observations, however, show that 300,000 years after the Big Bang, matter and radiation were distributed with variations in intensity of no more than a few parts in 10,000. Prior to that era, the universe was so hot that photons of energy were continually absorbed and re-emitted by matter, preventing stable atoms from forming. No lumps of matter could survive in the turmoil of the early universe. Hence, galaxies must have condensed *after* this epoch, when matter and radiation went their separate ways.

Thus the paradox: Galaxies form in clumps. Yet we know from COBE that the universe started out smooth and featureless. At the present time, astronomers cannot explain how the universe got from one state to the other because there doesn't appear to be enough time between the COBE era and the quasar/galaxy era for gravity to have collected matter into the complex clusters seen today. It is a problem serious enough to prompt a few researchers to suggest there is something wrong with the Big Bang theory.

The Three Pillars of Cosmology

Despite this so-called "structure problem," most astronomers are reluctant to abandon the Big Bang. The evidence supporting the theory, they say, is extensive and compelling enough to stand on its own. Astronomers cite three observations as the underpinnings of the Big Bang cosmology.

The first was made in the 1920s, when astronomers noted that the spectral lines of most galaxies (the lines arising from the emission or absorption of radiation by certain atoms at fixed wavelengths) appeared shifted toward the red end of the spectrum, a shift attributed to the galaxies apparently speeding away from us.

Then, in 1929, American astronomer Edwin Hubble discovered that a galaxy's recession velocity is directly proportional to its distance. Astronomers concluded that the universe is expanding and that

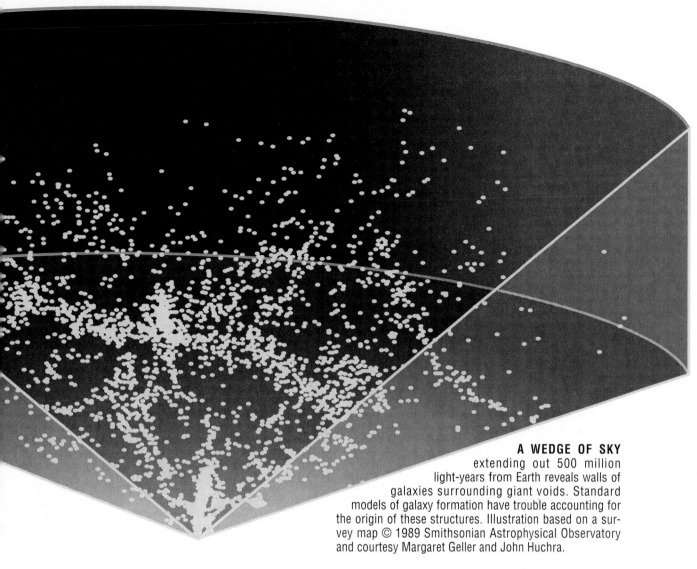

A WEDGE OF SKY extending out 500 million light-years from Earth reveals walls of galaxies surrounding giant voids. Standard models of galaxy formation have trouble accounting for the origin of these structures. Illustration based on a survey map © 1989 Smithsonian Astrophysical Observatory and courtesy Margaret Geller and John Huchra.

space between the galaxies is steadily increasing like the space between raisins in a cinnamon roll as it rises. For the universe to be expanding, however, a propulsive force of unimaginable magnitude — a Big Bang — must have set matter on its runaway course.

The second fundamental observation supporting a Big Bang is the measurement of the total abundance of light elements in the universe. This is akin to analyzing the ingredients of a cake after it's been baked. You know that for the cake to have the consistency, flavor, and structure that it has, certain substances must be present. From an analysis of samples of the cake you can derive a ratio of one ingredient to another, which can then be applied to the entire body of the cake.

Cosmologically speaking, the elements present in the early universe can be sampled by looking at the most ancient components of the universe — the oldest stars (usually found in globular clusters). The ratio of helium to hydrogen in these stars is exactly what the Big Bang theory predicts should have been forged by thermonuclear reactions 100 seconds after the primordial event. Though scientists have attempted to come up with other theories to account for the light-element abundances, none has accounted for them as well as the Big Bang.

Finally, we return to the cosmic background radiation, considered the most compelling argument for a Big Bang. The cosmic background temperature is 2.7 kelvins, and this radiation is uniform across the entire sky. Nowhere is the temperature higher or lower than this value. Slight variations would indicate hot spots in the background, the lumpy seedlings for the first clusters of galaxies. So far, none have been detected. This is exactly what the Big Bang model predicts.

Cracks in the Pillars?

But it is here that problems arise. For despite the three observational pillars, the Big Bang theory fails to fulfill the requirement for a complete cosmology: it does not explain it *all*. Everything in existence. Everywhere. The Big Bang theory explains only the first few moments of the universe; it stops short of explaining how galaxies and stars materialized in the first one billion years of the universe. This leads to the unsettling conclusion that, though astronomers believe they know how the universe began, they don't know how it got to its present state. Like the missing 18 minutes from the Watergate tapes, there is a narrow but crucial gap in the transcripts of the universe.

Filling the gap requires a theory that links the Big

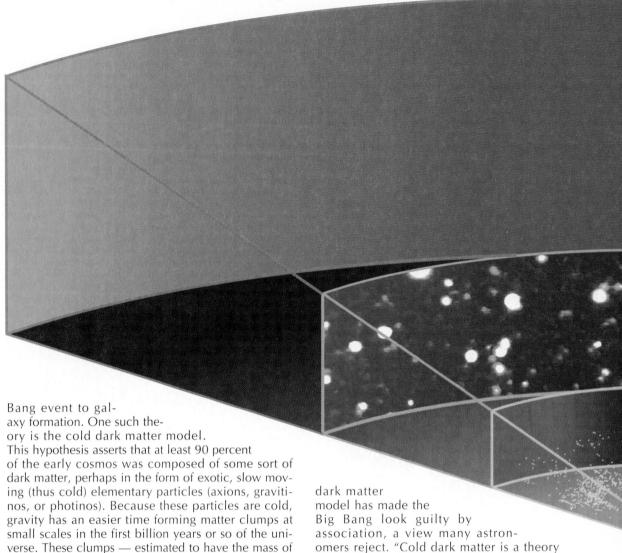

Bang event to gal-
axy formation. One such theory is the cold dark matter model.
This hypothesis asserts that at least 90 percent
of the early cosmos was composed of some sort of
dark matter, perhaps in the form of exotic, slow moving (thus cold) elementary particles (axions, gravitinos, or photinos). Because these particles are cold,
gravity has an easier time forming matter clumps at
small scales in the first billion years or so of the universe. These clumps — estimated to have the mass of
a typical globular cluster — gravitationally attract
more matter until galaxy-sized masses are built up.
The galaxies themselves then collect into the elabo-
rate clusters we see today. This is known as the "bottom-up" model of galaxy formation.

Unfortunately, cold dark matter, exotic or otherwise, has yet to be detected, save indirectly by what
are thought to be its gravitational effects on normal
matter, specifically the motions of galaxies in clusters
and the rotations of spiral galaxies. Astronomers have
yet to determine what form this dark matter takes.
(See "Shedding Light on Dark Matter," February 1992
ASTRONOMY.)

Compounding the problem, even the most gener-
ous cold dark matter models cannot account for the
astonishing dimensions of large-scale structures that
are turning up in galaxy surveys today — structures
that span nearly a billion light-years.

Is the Big Bang in trouble? Most astronomers think
not. They view the Big Bang and galaxy formation as
separate events. "I don't think anything that's happened recently has changed the majority opinion that
the Big Bang is sound," says Ethan Vishniac of the University of Texas at Austin. "I would say the current situ-
ation is that we have no believable theory of galaxy
formation. What has happened in the last year or so is
that . . . the cold dark matter theory has collapsed."

For some astronomers, the downfall of the cold
dark matter
model has made the
Big Bang look guilty by
association, a view many astronomers reject. "Cold dark matter is a theory
about the beginning of the formation of structure,"
says Princeton University cosmologist James Peebles. "The Big Bang is something quite different. The
problem is that the press has conflated the two.
They take the discussions about the beginnings of
structure and confuse them with the beginning of
the universe."

Taking issue with Vishniac, Peebles, and practi-
cally the entire astronomical community is Anthony
L. Peratt, a physicist and cosmologist at Los Alamos
National Laboratory. Rather than propose a better
theory of galaxy formation, Peratt suggests scrapping
the Big Bang altogether.

"The Big Bang theorists attempt to decouple
themselves from the problem of galaxy formation,"
says Peratt. "But one must ask what kind of cosmology is it that cannot account for the galaxies and
stars we observe? This was a basic requirement for
all previous cosmologies, from the time of the Ioni-
ans to Newton. Big bang cosmologists have strug-
gled mightily to find a mechanism for galaxy formation in spite of their arguments that this is not their
problem."

Peratt belongs to a group of scientists who sub-
scribe to an alternative cosmogenesis theory called
"plasma cosmology," which argues that the laws of
electromagnetism, rather than gravity, dominate and
shape the universe.

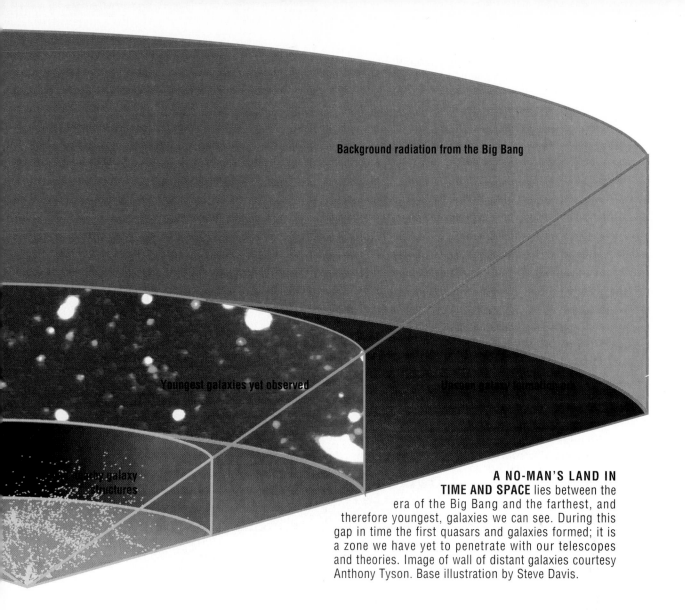

Background radiation from the Big Bang

Youngest galaxies yet observed

Unseen galaxy formation

Nearby galaxy structures

A NO-MAN'S LAND IN TIME AND SPACE lies between the era of the Big Bang and the farthest, and therefore youngest, galaxies we can see. During this gap in time the first quasars and galaxies formed; it is a zone we have yet to penetrate with our telescopes and theories. Image of wall of distant galaxies courtesy Anthony Tyson. Base illustration by Steve Davis.

An Alternate Universe

Plasma is a state of matter that resembles a gas, except that instead of consisting of electrically neutral atoms, plasma contains charged particles — electrons torn from atoms in a gas, leaving a cloud of negatively charged free electrons and positively charged ions. Although affected by gravity, these charged particles can also be accelerated by electromagnetic fields.

Most of the visible mass of the universe is plasma. On Earth, the most dramatic forms are lightning and aurorae. In space, stars are gravitationally bound plasmas. On larger scales, plasmas have been detected at the galactic center and in the double lobes of radio galaxies. The largest plasma structure was discovered in 1989, when faint radio emission was detected between two superclusters of galaxies, indicating that the superclusters are embedded in a warm plasma.

The plasma cosmological model was first proposed by Hannes Alfvén, of the Royal Institute of Technology in Stockholm, Sweden. In 1970, Alfvén was awarded the Nobel Prize in physics for his work in solar magnetohydrodynamics — the movement of plasma in magnetic fields. In separate work, Alfvén reasoned that sheets of electric currents crisscross the universe. These electromagnetic fields and not gravity, plasma cosmologists suggest, are the major movers of matter on the grandest scales. Interaction with electromagnetic fields, Peratt says, enables plasmas to exhibit complex structure and motion that far exceed what's possible with gravity alone.

Plasma proponents have conducted computer simulations involving tens of millions of test particles (used to represent galaxies) to demonstrate that a fundamentally uniform plasma will eventually break up into cellular structures that look remarkably similar to the large-scale structures seen today.

The cellular structures would, in turn, act to scatter radiation. Plasma cosmologists theorize that what we detect as the cosmic background actually results when local fields and currents scatter microwave radiation from pervasive plasmas like a dense fog scatters a car's headlights.

But what about the successful prediction of the abundance of light elements proclaimed by Big Bang proponents? "We have always questioned just how well known the abundances of light elements are in the universe," says Peratt. "And if the universe is cellular and filamentary, what are the abundances of the elements within and without the filaments?" Peratt cites a recent measurement of light elements in galaxies that indicates a helium abundance *lower* than that allowed by the standard Big Bang model. A lower assumed density of the universe would produce such a lower helium abundance, but that would also mean the standard Big Bang model does not satisfy the primordial abundance constraints for the other light elements for any density. Interestingly, such low helium abundances were found by other researchers in 1988.

As for the "missing link" between the Big Bang and the first galaxies, plasma physicists claim there is no missing link because there never was a matter-forming epoch. As Peratt writes in an article in the January/February 1990 issue of *The Sciences*, "There is no expansion, and there need not be any final crunch. Unlike the universe envisioned in the Big Bang model, the plasma universe evolves without beginning and without end: it is indefinitely ancient and has an indefinite lifetime in store."

Peratt's plasma cosmology isn't the only alternative idea. The steady state model still has a few supporters. Steady state theorists claim that new galaxies are constantly forming to replace old galaxies at a rate determined by the expansion rate of the universe. Supporters also argue that the cosmic microwave background is actually normal radiation from stars, radiation that has been scattered by metallic "whiskers" sown throughout the interstellar medium in supernovae ejecta.

But don't confuse plasma cosmology with steady state theories, explains Peratt. Steady state invokes a perfect cosmological principle in which the universe ap-pears the same at all points at all times — past, present, and future. In contrast, the plasma universe does evolve with time. According to Peratt, "The plasma universe model has nothing to do with the steady state model. Both the Big Bang and the steady state espouse gravity as the sculptor of the universe."

Paradox Lost?

Among the supporters of alternative cosmologies, there are few agreements except, perhaps, in

their enthusiasm for refuting conventional cosmology. Each has a response to the arguments posed by Big Bang cosmologists.

Mainstream astronomers counter these attacks by pointing out that alternative theories like the plasma and steady state cosmologies don't come close to the predictive power of the Big Bang. It has withstood three-quarters of a century of scrutiny, they say. It will take more than a few stabs to topple it.

"We don't really have a complete theory of planet formation but we wouldn't use that to question that the Sun formed from contracting interstellar gas," says Lennox Cowie of the University of Hawaii. Cowie has made a specialty of probing the distant universe for primeval galaxies. Recently, he and a team of astronomers completed a deep infrared imaging survey that turned up thousands of faint dwarf galaxies approximately one hundredth the size of our Galaxy lying at moderate distances. These faint galaxies are likely the most populous type of galaxy in the universe. But more importantly, he adds, "They probably contain at least as much

FROM SMOOTH TO FOAMY, a series of computer simulations shows the transition from the uniform structure that arose out of the Big Bang (bottom left) to the filaments and voids of the universe today (top right). Current models of this process are now under suspicion. Images of "cold dark matter" simulation courtesy James Gelb, MIT.

mass as normal galaxies." A universe filled with dwarf galaxies may account for a great deal of the missing mass in the universe, mass predicted by the cold dark matter model. And if you have more mass, you can account for more structure. So perhaps cold dark matter is still lurking out there somewhere, waiting to be found.

The Big Bang theory could also be strengthened by looking at galaxy formation in a new way. In December 1991 a team of radio astronomers at the Very Large Array in New Mexico announced the discovery of a large mass of hydrogen gas near the edge of the observable universe. According to the "top-down" theory of galaxy formation, galaxy clusters formed when giant primordial gas clouds broke apart into smaller gas clouds and not when smaller clouds came together as in the "bottom-up" scenario. A refined top-down idea may explain how large structures can form in a relatively short time span, solving the current problem. The new discovery may be the first evidence for the process.

The bottom line is that the majority of astronomers are reluctant to toss out the Big Bang theory because of an inability to explain how galaxies form. "The problem of galaxy formation is one that depends on rather poorly understood physics and also on some rather shaky assumptions," Cowie says. "The Big Bang theory on the other hand predicts both the light element abundances and the microwave background. I think until we get a much better understanding of the history of galaxy formation it would be very presumptuous to question the underlying cosmological model."

Princeton University's James Peebles agrees. In fact, in a paper published in the August 29, 1991 issue of the British journal *Nature*, Peebles, along with fellow Princeton cosmologist Edwin Turner, and

David Schramm and Richard Kron of the University of Chicago, defend the Big Bang. They point out that plasma cosmology can't explain why the cosmic background radiation is so uniform over the entire sky. And that is just one of many problems they see with the alternative cosmologies.

As Peebles puts it, "Time and again people have said, It's difficult to think of a theory of galaxy formation that can be consistent with all observations. So perhaps that difficulty means that the framework within which we're working — the Big Bang — is wrong.' And time and again that inference has been shown to be incorrect.

"By being a little more clever in the invention of theories," Peebles says, "a person can find a way around the problems and continue to look for a way to build within the Big Bang framework."

Peratt, however, remains steadfast. "Plasma cosmologists don't feel we are 'closing in on creation. . . .' In fact, if our findings are anywhere near right, the universe is much bigger and older than previously imagined."

The conflict between the differing views of the universe presents a classic example of an established theory undergoing a revolution. Though the Big Bang model may have endured for 75 years, history is replete with erroneous scientific beliefs that survived centuries, as well as some "crackpot" theories that later became accepted (continental drift, for example).

Is the Big Bang theory wrong? It is certainly under attack. Scientists are probing the Achilles' heel of this venerable cosmology in a debate sparked by what new telescopes probing to the edge of the universe have shown us.

But debate is what science thrives on. Without it, we would be like a people starving to death in a land of sacred cows. □

Jeff Kanipe, a former associate editor of ASTRONOMY, is now the editor of Star Date *magazine, published by McDonald Observatory and the University of Texas at Austin.*